Continuous Time Dynamical Systems

State Estimation and Optimal Control with Orthogonal Functions

B. M. Mohan
S. K. Kar

CRC Press
Taylor & Francis Group
Boca Raton London New York

CRC Press is an imprint of the
Taylor & Francis Group, an **informa** business

CRC Press
Taylor & Francis Group
6000 Broken Sound Parkway NW, Suite 300
Boca Raton, FL 33487-2742

First issued in paperback 2017

© 2013 by Taylor & Francis Group, LLC
CRC Press is an imprint of Taylor & Francis Group, an Informa business

No claim to original U.S. Government works

ISBN-13: 978-1-4665-1729-5 (hbk)
ISBN-13: 978-1-138-07358-6 (pbk)

Library of Congress Cataloging-in-Publication Data

Mohan, B. M. (Bosukonda Murali)
    Continuous time dynamical systems : state estimation and optimal control with orthogonal functions / B.M. Mohan, S.K. Kar.
        p. cm.
    Includes bibliographical references and index.
    ISBN 978-1-4665-1729-5 (hardback)
    1. Functions, Orthogonal. 2. Differentiable dynamical systems--Automatic control--Mathematics. 3. Mathematical optimization. I. Kar, S. K. II. Title.

QA404.5.M64 2012
515'.55--dc23   2012028470

Visit the Taylor & Francis Web site at
http://www.taylorandfrancis.com

and the CRC Press Web site at
http://www.crcpress.com

Dedication

To my wife, Vijaya Lakshmi,

my son, Divyamsh &

my daughter, Tejaswini

B. M. Mohan

To my parents, Satya Ranjan & Manjushree,

my sister, Sanjeeta,

my wife, Swastika &

my daughters, Santika & the late Ayushi

S. K. Kar

Contents

viii

List of Abbreviations

BPFs	Block pulse functions
CP1s	Chebyshev polynomials of the first kind
CP2s	Chebyshev polynomials of the second kind
GOPs	General orthogonal polynomials
HFs	Haar functions
HPs	Hermite polynomials
ISE	Integral square error
JPs	Jacobi polynomials
LaPs	Laguerre polynomials
LPs	Legendre polynomials
LQE	Linear-quadratic-estimator
LQG	Linear-quadratic-Gaussian
LQR	Linear-quadratic-regulator
MIMO	Multi-input multi-output
NSR	Noise-to-signal ratio
OFs	Orthogonal functions
SCFs	Sine-cosine functions
SCP1s	Shifted Chebyshev polynomials of the first kind
SCP2s	Shifted Chebyshev polynomials of the second kind

SJPs   Shifted Jacobi polynomials

SLPs   Shifted Legendre polynomials

TPBV   Two-point boundary-value

WFs   Walsh functions

List of Figures

Preface

Orthogonal functions can be broadly classified into three families: the piecewise constant family, the polynomial family, and the sine-cosine family. Each family is classified into different classes of basis functions. Classes such as block-pulse functions, Haar functions and Walsh functions belong to the piecewise constant family while Chebyshev polynomials of the first kind and the second kind, Legendre polynomials, Laguerre polynomials, Hermite polynomials etc. belong to the polynomial family. The techniques of reducing the (differential, integral) calculus of continuous-time dynamical systems to an attractive algebra, approximate in the sense of least squares, emerged in the early 1970s, first with piecewise constant family functions. In subsequent years the use of polynomial family and sine-cosine functions has been demonstrated.

It all began with the Walsh functions application to optimal control problems in 1975, and the block-pulse functions application to state estimation problems in 1984. Since then, block-pulse functions, Legendre polynomials and other classes of basis functions have been applied extensively for state estimation of linear time-invariant systems, and optimal control of varieties of systems such as linear/bilinear/nonlinear, time-invariant/time-varying, and delay-free/delay systems. All these studies show that block-pulse functions and Legendre polynomials have definite computational advantages in comparison to other classes of basis functions, as they can easily lead to recursive algorithms.

The work of the authors and the several interesting contributions of other researchers form the core of this book. It is perhaps the first book devoted exclusively to the application of block-pulse functions and Legendre polynomials to state estimation and optimal control of continuous-time dynamical systems. It is hoped that this book will be of interest to control students and engineers.

B. M. Mohan

S. K. Kar

Acknowledgements

The authors thank the authorities of the Indian Institute of Technology, Kharagpur and the Institute of Technical Education Research, Sikha 'O' Anusandhan University, Bhubaneswar for the facilities and the conducive atmosphere provided for the research and preparation of this book. They would like to express their gratitude to Professor G. P. Rao and Professor K. B. Datta, both retired from IIT Kharagpur, for their helpful discussions, keen interest and kind encouragement. They are very much indebted to their family members for their love, understanding and support during the years of their research and the period of preparation of this book. Last but not the least, they also thank Dr. Gagandeep Singh, Commissioning Editor, Kari Budyk, Senior Project Coordinator and Karen Simon, Project Editor, of CRC Press, Taylor & Francis Group, for their continued perseverance and encouragement for making this book a reality.

About the Authors

B. M. Mohan received his bachelor's degree in electrical engineering from Osmania University in 1982, a master's degree in electrical engineering (with control systems specialization) from Andhra University in 1985, and his doctoral degree in engineering from the Indian Institute of Technology, Kharagpur in 1989. From July 1989 to October 1990 he was on the faculty of the Electrical and Electronics Engineering Department, Regional Engineering College (now called the National Institute of Technology), Tiruchirapalli. Since November 1990 he has been on the faculty of the Electrical Engineering Department, Indian Institute of Technology, Kharagpur, where he is currently a professor. He was a visiting professor in the Department of Electrical Engineering–Systems, the University of Southern California, Los Angeles, California in 2006.

Dr. Mohan's research interests include identification, analysis, and control of dynamical systems using fuzzy logic and orthogonal functions. He co-authored the research monograph *Orthogonal*

Functions in Systems and Control (World Scientific, Singapore, 1995), and several papers in international journals and conferences. He is a member of the Asian Control Association, a life member of the Systems Society of India, a senior member of the IEEE, and a life fellow of the Institution of Engineers (India). He is the associate editor of the *International Journal of Automation & Control* (IJAAC); an editorial board member of the *International Journal of Mathematics and Engineering with Computers* (IJMAEC), the *Journal of Control Engineering & Technology* (JCET), the *International Journal of Engineering Science, Advanced Computing & Bio-Technology* (IJESACBT), and the *International Journal of Control, Automation & Systems* (IJCAS). He is also the editor of the *Journal of Electrical & Control Engineering* (JECE), and *Current Trends in Systems & Control Engineering* (CTSCE). He was the chair of the Control, Robotics & Motion Control track, ICIIS 2008.

Sanjeeb Kumar Kar received his B.Tech in electrical engineering from the College of Engineering & Technology (CET),

OUAT, Bhubaneswar in 1993, and his M. Tech in control systems engineering and Ph.D from the Indian Institute of Technology, Kharagpur, in 2004 and 2011, respectively. From March 2004 to February 2005 he worked as an apprentice engineer on the Odisha State Electricity Board, and from April 1995 to March 1997 as assistant manager in Pal Textile, Balasore. In August 1997 he joined the faculty in electrical engineering, CET, OUAT, Bhubaneswar and continued there until October 2008. Since November 2008, he has been on the faculty of the Electrical Engineering Department, the Institute of Technical Education & Research (ITER), and Siksha 'O' Anusandhan (SOA) University, Bhubaneswar, where he is currently working as an associate professor and head of the department.

Chapter 1

Introduction

The optimal control problem is introduced first. Then optimal control problems of varieties of systems and their solution via different classes of orthogonal functions (OFs) are discussed in the literature review section. The objectives and contributions of the book are stated. The organization of the book is given in the last section.

1.1 Optimal Control Problem

A particular type of system design problem is the problem of "controlling" a system. The translation of control system design objectives into the mathematical language gives rise to the control problem. The essential elements of the control problem are

- A desired output of the system.

- A set of admissible inputs or "controls."

- A performance or cost functional which measures the effectiveness of a given "control action."

The objective of the system is often translated into a requirement on the output. Since "control" signals in physical systems are usually obtained from equipment which can provide only a limited amount of force or energy, constraints are imposed upon the inputs to the system. These constraints lead to a set of admissible inputs.

Frequently the desired objectives can be attained by many admissible inputs, so the engineer seeks a measure of performance or cost of control which will allow him/her to choose the "best" input. The choice of a mathematical performance functional is a subjective matter. Moreover, the cost functional will depend upon the desired behaviour of the system. Most of the time, the cost functional chosen will depend upon the input and the pertinent system variables. When a cost functional has been decided upon, the engineer formulates his/her control problem as follows:

Determine the admissible inputs which generate the desired output and which optimize the chosen performance measure.

At this point, optimal control theory enters the picture to aid the engineer in finding a solution to his/her control problem. Such a solution, when it exists, is called an optimal control. Optimal control deals with the problem of finding a control law for a given system such that a certain optimality criterion is achieved. An optimal control is a set of differential equations describing the paths of the control variables that minimize the cost functional. The theory of optimal control is concerned with operating a dynamic system at minimum cost. The case where the system dynamics are described by a set of linear differential equations and the cost is described by a quadratic functional is called a linear quadratic problem. One of the main results in the theory is that the solution is provided by the linear-quadratic-regulator (LQR).

In this book different classes of systems are considered with quadratic performance criteria and an attempt is made to find the optimal control law for each class of systems using OFs that can optimize the given performance criteria.

1.2 Historical Perspective

The Legendre polynomials (LPs) originated from determining the force of attraction exerted by solids of revolution [5], and their orthogonal properties were established by Adrian Marie Legendre during 1784–1790. There is another family of OFs known as piecewise constant basis functions whose functional values are constant within any subinterval of time period. There is a class of complete OFs known as block-pulse functions (BPFs) which is more popular and elegant in the areas of parameter estimation, analysis and control.

Control of linear systems by minimizing a quadratic performance index gives rise to a time-varying gain for the linear state feedback, and this gain is obtained by solving a matrix Riccati differential equation [3]. Probably, Chen and Hsiao, 1975, were the first who applied a class of piecewise constant OFs, i.e. Walsh functions (WFs), obtained a numerical solution of the matrix Riccati equation [6] and found the time-varying gain. Then many researchers started investigating the problems of identification, analysis and control using different classes of OFs. The operational matrix for integration of BPFs was derived [8]. Moreover, it was shown that BPFs are more fundamental than WFs and the structure of the integration operational matrix of BPFs is simpler than that of WFs. In [9] it was shown that optimal control problems could be solved using BPFs with minimal computational effort.

In the last three and a half decades, the OF approach was successfully applied to study varieties of problems in systems and control [22, 57, 59, 61]. The key feature of OFs is that it converts differential or integral equations into algebraic equations in the sense of least squares. So this approach became quite popular computationally as the dynamical equations of a system can be converted into a set of algebraic equations whose solution leads to the solution of the problem.

Before going into the details of the optimal control problem, we first take a look at the problem of estimation of state variables as the state estimation plays an important role in the context of state feedback control. State feedback control system design requires the knowledge of the state vector of the plant. Sometimes, no state variables or only a few state variables are available for measurement. In such cases an observer, either full order or reduced order depending on the situation, is incorporated to estimate the unknown state variables if the plant is observable. In general, it is estimated using the Luenberger observer [1]. But the Luenberger observer produces erroneous estimates in a noisy environment unless the measurement noise is filtered out. Interestingly, the OF approach has an inherent filtering property [59] as it involves an integration process which has the smoothing effect. As it appears from the literature, two attempts have been made on the state estimation problem by using two different classes of OFs, i.e. BPFs [23] and shifted Chebyshev polynomials of first kind (SCP1s)[39] so far. It is observed that the BPF approach [23] is purely recursive and uses multiple integration. The number of integrations increases as the order of the system increases, i.e for an nth order system the state equation has to be integrated n times, which is computationally not attractive.

Next, coming to the SCP1 approach [39], integration operational matrix of SCP1s is less sparse than that of shifted Legendre polynomials (SLPs). So if we use SLPs to develop algorithms, it will obviously be more elegant computationally. Moreover, state estimation cannot be done via SCP1s in a noisy environment.

The problem of optimal control incorporating observers has been successfully studied via different classes of OFs, namely BPFs [19], SLPs [30, 47], shifted Jacobi polynomials (SJPs) [33], general orthogonal polynomials (GOPs) [41], sine-cosine functions (SCFs) [44, 47], SCP1s [25, 47], shifted Chebyshev polynomials of the second kind (SCP2s) [47] and single-term Walsh series [52]. The approach followed in [25, 30, 33, 41, 44, 47] is nonrecursive, while it is recursive in [19, 52], making the approach in [25, 30, 33, 41, 44, 47] computationally not attractive.

Synthesis of optimal control laws for deterministic systems described by integro-differential equations has been investigated [11] via the dynamic programming approach. Subsequently, this problem has been studied via BPFs [37], SLPs [34, 43], and SCP1s [42].

The linear-quadratic-Gaussian (LQG) control problem [4] concerns linear systems disturbed by additive white Gaussian noise, incomplete state information and quadratic costs. The LQG controller is simply the combination of a linear-quadratic-estimator (LQE), i.e. Kalman filter, and an LQR. The separation principle guarantees that these can be designed and computed independently. In [35] the solution of the LQG control design problem has been obtained by employing GOPs. By using the GOPs the nonlinear Riccati differential equations have been reduced to nonlinear algebraic equations. The set of nonlinear algebraic equations has been solved to get the solutions. The above approach is neither simple nor elegant computationally, as nonlinear equations are involved.

Singular systems have been of considerable importance as they are often encountered in many areas. Singular systems arise naturally in describing large-scale systems [53]; examples occur in power and interconnected systems. In general, an interconnection of state variable subsystems is conveniently described as a singular system. The singular system is called generalized state-space system, implicit system, semi-state system, or descriptor system. Optimal control of singular systems has been discussed in [15] and [18]. In [38] the necessary conditions for the existence of optimal control have been derived. It has been shown that the optimal control design problem reduces to a two-point boundary-value (TPBV) problem for the determination of the optimal state trajectory. The single-term Walsh series method [56] has been applied to study the optimal control problem of singular systems. In [62] SLPs were used to solve the same problem. However, this approach is nonrecursive in nature. The Haar wavelet approach [68] has been presented to study the optimal control problem of linear singularly perturbed systems. In the recent times, SCFs [71], SCP1s [74] and Legendre wavelets [82] have been applied for solving the optimal control problem of singular systems. These approaches are again nonrecursive.

Time-delay systems are those systems in which time delays exist between the application of input or control to the system and its resulting effect on it. They arise either as a result of inherent delays in the components of the system or as a deliberate introduction of time delays into the system for control purposes. Examples of such systems are electronic systems, mechanical systems, biological systems, environmental systems, metallurgical systems, chemical systems, etc. A few practical examples [46] are controlling the speed of a steam engine running an electric power generator under varying load conditions, and control of room temperature, a cold rolling mill, spaceship, hydraulic system, etc.

As it appears from the literature, extensive work was done on the problem of optimal control of linear continuous-time dynamical systems containing time delays. Palanisamy and Rao [20] appear to be the first to study the optimal control problem via WFs. They considered time-invariant systems with one delay in state and one delay in control. In [24] time-varying systems containing one delay in state and one delay in control were considered, and optimal control problems of such systems was studied via BPFs. Solutions obtained in [20, 24] are piecewise constant. In order to obtain smooth solution, SCP1s [27] were used to study time-invariant systems with a delay in state only. In [31] time-varying systems with multiple delays in state and control have been studied via SLPs. The problem considered in [20] was investigated again by applying SLPs [36]. In [48] the problem investigated in [24] was again solved by approximating the time-delay terms via Pade approximation and using GOPs. Similarly, in [58] the problem considered in [27] was studied again via SLPs. Linear Legendre multiwavelets [81] were used to solve the time-varying systems having a delay in state only.

In recent years, people have come up with a new idea of defining hybrid functions (with BPFs and any class of polynomial functions) and utilizing the same for studying problems in Systems and Control. The so-called hybrid functions approach was first introduced in [67] to study the optimal control problem of time-varying systems with a time-delay in state only. Subsequently, this approach was extended to time-invariant systems [72] studied in [27], delay systems containing reverse time terms [73], and time-varying systems [83] considered in [67]. In [84] general Legendre wavelets were used to solve the optimal control problem of time-varying singular systems having a delay in state only.

Looking at the historical developments on solving optimal control problem of nonlinear systems via the OF approach, we find that not much work has been reported. Lee and Chang [40] appear to be the first to study the optimal control problem of nonlinear systems using GOPs. For this, they introduced a nonlinear operational matrix of GOPs. Though their work is fundamental and significant, it is not attractive computationally, as it involves the Kronecker product [12] operation. Chebyshev polynomials of first kind (CP1s) were used [49] for solving nonlinear optimal control problems. In [54] a general framework for nonlinear optimal control problems was developed by employing BPFs.

A BPF approach to hierarchical control of linear time-varying large-scale systems has been reported [45]. The resulting solutions are found to be piecewise constant with minimal mean-square-error.

1.3   Organisation of the Book

The book consists of eleven chapters in all. They are

Chapter 1: Introduction.

Chapter 2: Orthogonal Functions and Their Properties; in this chapter brief discussion on OFs and their classification are given. BPFs and their properties are given in Section 2.2.

LPs are presented in Section 2.3. SLPs and their properties are discussed in Section 2.4. A nonlinear operational matrix using LPs and BPFs is introduced in Section 2.5.

Chapter 3: State Estimation; estimation of unknown state variables is done using full order observer, both in noisy and noise free environments. The algorithm in [23] is modified to deal with nonzero initial state vector.

Chapter 4: Linear Optimal Control Systems Incorporating Observers; using a reduced order observer the unknown state variables are estimated and the optimal control law is obtained.

Chapter 5: Optimal Control of Systems Described by Integro-Differential Equations; a unique approach via BPFs or SLPs is presented in Section 5.2 to solve the optimal control problem of linear systems described by integro-differential equations.

Chapter 6: Linear-Quadratic-Gaussian Control; the LQG control design problem is discussed in Section 6.2. Solutions are given in Sections 6.3 and 6.4 which contain a unified approach and two recursive algorithms.

Chapter 7: Optimal Control of Singular Systems; in Section 7.2 a singular system problem is solved as an initial value problem under some conditions and two recursive algorithms are given. A more generalized solution to the problem is given in Section 7.3.

Chapter 8: Optimal Control of Time-Delay Systems; an approach to compute optimal control law of linear time-varying multi-delay dynamic systems with a quadratic performance index is discussed in Section 8.2. Then application of the approach to

- Time-invariant systems
- Delay free systems
- Singular systems with delays

is given in Subsections 8.2.3, 8.2.4 and 8.2.5. In Section 8.3 optimal control of delay systems with reverse time terms is discussed and algorithms are presented.

Chapter 9: Optimal Control of Nonlinear Systems; a new method for computing optimal control law for nonlinear systems is given by employing LPs and BPFs.

Chapter 10: Hierarchical Control of Linear Systems; a BPF method of hierarchical control of linear time-invariant/time-varying large scale systems is presented.

Chapter 11: Epilogue; the last chapter concludes the book with the scope for further research.

Chapter 2

Orthogonal Functions and Their Properties

In this chapter a brief discussion on the history of OFs, their classification and their developments is given. The useful properties of BPFs and SLPs are given in detail. Lastly, the rationale for choosing BPFs and SLPs among all OFs, for the study of state estimation and optimal control problems, is stated.

2.1   Introduction

The LPs, originated from determining the force of attraction exerted by solids of revolution [5], and their orthogonal properties were established by Adrian Marie Legendre (1752–1833) during 1784–1790. In 1807, Joseph Fourier (1768–1830), while solving partial differential equations encountered in connection with conduction of heat in a rod, discovered that the solution can be expressed as a series of exponentially weighted sine functions. Later he extended this idea to represent any arbitrary function as an infinite sum of sine and cosine functions. Pafnuty Lvovich Chebyshev (1821–1894), a Russian mathematician, observed that of all polynomials approximation of an arbitrary function in the interval $-1 \leq x \leq 1$, the one that minimizes the maximum error is a linear combination of polynomials known as the Chebyshev polynomials [5]. Hermite polynomials (HPs) were introduced [5] in 1864 by the mathematician Charles Hermite (1822–1905). Edmond Laguerre (1834–1866) discovered Laguerre polynomials (LaPs) while converting a divergent power series into a convergent continued fraction. There is another family of OFs known as piecewise constant basis functions whose functional values are constant within any subinterval of time period. In this family, in 1910, Haar functions (HFs) were proposed by the Hungarian mathematician Alfred Haar. Subsequently in 1923, Joseph Leonard Walsh, an American mathematician, introduced another set of complete OFs called WFs. There is a class of complete OFs known as BPFs which have become popular in solving the problems of parameter estimation, analysis and control.

In general, depending upon their nature, OFs may be broadly classified in three categories:

- Piecewise Constant Orthogonal Functions,

- Orthogonal Polynomials, and

- Sine-Cosine (Fourier) Functions.

The BPFs, HFs and WFs belong to the family of piecewise constant orthogonal functions while LPs, LaPs, HPs, CP1s, Chebyshev polynomials of the second kind (CP2s) and Jacobi polynomials (JPs) belong to the family of orthogonal polynomials. Since each class of these functions forms a basis for the series expansion of a square-integrable function, OFs are commonly referred to as basis functions.

A set of functions $\{\phi_i(t)\}$, $i = 0, 1, \ldots, m-1$ is said to be orthogonal on $t_0 \leq t \leq t_f$ with respect to a nonnegative weighting function $w(t)$ if

$$\int_{t_0}^{t_f} w(t)\phi_i(t)\phi_j(t)dt = \begin{cases} 0, & i \neq j \\ \gamma_i, & i = j \end{cases} \tag{2.1}$$

where γ_i is a non-zero positive constant and $w(t)$ is not zero throughout the interval. If $\gamma_i = 1$, the set $\{\phi_i(t)\}$ is known as an orthonormal set. The class of functions satisfying the condition

$$\int_{t_0}^{t_f} w(t)f^2(t)dt < \infty$$

is known as square integrable over the interval $[t_0, t_f]$ and can be expressed in terms of OFs.

A set of OFs $\{\phi_0, \phi_1, \ldots, \phi_r, \ldots\}$ is said to be complete if for any continuous square integrable function $f(t)$, the following integral square error (ISE) tends to zero as m tends to infinity.

$$\text{ISE} = \int_{t_0}^{t_f} w(t)\left[f(t) - \sum_{i=0}^{m-1} f_i\phi_i(t)\right]^2 dt \tag{2.2}$$

Then Parseval's relation or the relation of completeness is given by

$$\int_{t_0}^{t_f} w(t)f^2(t)dt = \sum_{i=0}^{\infty} \gamma_i f_i^2 \tag{2.3}$$

The OF approach became quite popular numerically and computationally as it converts calculus (differential or integral) into algebra in the sense of least squares, i.e. dynamical equations of a system can be converted into a set of algebraic equations whose solution simply leads to the solution of dynamical equations. In the last three and half decades, OF approach was successfully applied to study varieties of problems in Systems and Control [22, 57, 59, 61]. We consider two classes of OFs, namely BPFs and LPs (or SLPs) among all classes of OFs, in the study of state estimation and optimal control problems. Their properties are discussed here.

2.2 Block-Pulse Functions (BPFs)

A set of m BPFs, orthogonal over $t \in [t_0, t_f)$, is defined [22] as

$$B_i(t) = \begin{cases} 1, & t_0 + iT \le t < t_0 + (i+1)T \\ 0, & \text{otherwise} \end{cases} \tag{2.4}$$

for $i = 0, 1, 2, \ldots, m-1$, where

$$T = \frac{t_f - t_0}{m}, \quad \text{the block-pulse width} \tag{2.5}$$

A set of four BPFs over $t \in [0, 1)$ is shown in Fig. 2.1. A square integrable function $f(t)$ on $t_0 \le t \le t_f$ can be approximately represented in terms of BPFs as

$$f(t) \approx \sum_{i=0}^{m-1} f_i B_i(t) = \mathbf{f}^T \boldsymbol{B}(t) \tag{2.6}$$

where

$$\mathbf{f} = \begin{bmatrix} f_0, & f_1, & \ldots, & f_{m-1} \end{bmatrix}^T \tag{2.7}$$

is an m − dimensional block-pulse spectrum of $f(t)$, and

$$\boldsymbol{B}(t) = \begin{bmatrix} B_0(t), & B_1(t), & \ldots, & B_{m-1}(t) \end{bmatrix}^T \tag{2.8}$$

an m − dimensional BPF vector. f_i in Eq. (2.6) is given by

$$f_i = \frac{1}{T} \int_{t_0+iT}^{t_0+(i+1)T} f(t)dt \tag{2.9}$$

which is the average value of $f(t)$ over $t_0 + iT \le t \le t_0 + (i+1)T$.

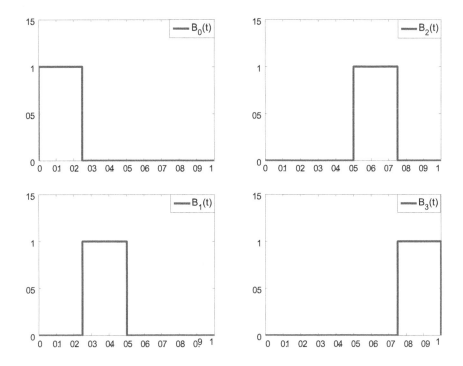

Figure 2.1: A set of four BPFs.

2.2.1   Integration of $B(t)$

Integrating $B(t)$ from t_0 to t and expressing the result in $m-$set of BPFs, we have

$$\int_{t_0}^{t} B(\tau)d\tau \approx H_f B(t) \qquad (2.10)$$

where

$$H_f = T \begin{bmatrix} \frac{1}{2} & 1 & 1 & 1 & \cdots & 1 \\ 0 & \frac{1}{2} & 1 & 1 & \cdots & 1 \\ 0 & 0 & \frac{1}{2} & 1 & \cdots & 1 \\ 0 & 0 & 0 & \frac{1}{2} & \cdots & 1 \\ \vdots & \vdots & \vdots & \vdots & & \vdots \\ 0 & 0 & 0 & 0 & \cdots & \frac{1}{2} \end{bmatrix} \qquad (2.11)$$

is called the operational matrix of forward integration [8] of BPFs and it is an $m \times m$ upper triangular matrix. Similarly if we integrate $\boldsymbol{B}(t)$ from t_f to t and express the result in $m-$set of BPFs, we have

$$\int_{t_f}^{t} \boldsymbol{B}(\tau)d\tau \approx H_b \boldsymbol{B}(t) \tag{2.12}$$

where

$$H_b = -T \begin{bmatrix} \frac{1}{2} & 0 & 0 & 0 & \cdots & 0 \\ 1 & \frac{1}{2} & 0 & 0 & \cdots & 0 \\ 1 & 1 & \frac{1}{2} & 0 & \cdots & 0 \\ 1 & 1 & 1 & \frac{1}{2} & \cdots & 0 \\ \vdots & \vdots & \vdots & \vdots & & \vdots \\ 1 & 1 & 1 & 1 & \cdots & \frac{1}{2} \end{bmatrix} = -H_f^T \tag{2.13}$$

is called the operational matrix of backward integration [16] of BPFs and it is an $m \times m$ lower triangular matrix.

2.2.2 Product of two BPFs

The product of two BPFs $B_i(t)$ and $B_j(t)$ can be expressed [22] as

$$B_i(t)B_j(t) = \begin{cases} 0 & \text{if } i \neq j \\ B_i(t) & \text{if } i = j \end{cases} \tag{2.14}$$

This is called disjoint property of BPFs.

2.2.3 Representation of $C(t)\mathbf{f}(t)$ in terms of BPFs

Let $C(t)$ be an $n \times n$ matrix and $\mathbf{f}(t)$ be an $n \times 1$ vector. Assume that the elements of $C(t)$ and $\mathbf{f}(t)$ are square integrable over $t_0 \leq t \leq t_f$. Then

$$C(t)\mathbf{f}(t) \simeq \sum_{i=0}^{m-1} C_i B_i(t) \sum_{j=0}^{m-1} \mathbf{f}_j B_j(t) = \sum_{i=0}^{m-1} C_i \mathbf{f}_i B_i(t) \tag{2.15}$$

Let

$$C(t)\mathbf{f}(t) = \mathbf{g}(t) \simeq \sum_{i=0}^{m-1} \mathbf{g}_i B_i(t) = G\mathbf{B}(t) \qquad (2.16)$$

where

$$G = \left[\begin{array}{cccc} \mathbf{g}_0, & \mathbf{g}_1, & \cdots, & \mathbf{g}_{m-1} \end{array} \right] \qquad (2.17)$$

Upon comparing Eqs. (2.15) and (2.16), we have [16]

$$\mathbf{g}_i = C_i \mathbf{f}_i \qquad (2.18)$$

2.2.4 Representation of a time-delay vector in BPFs

Assume that $\mathbf{f}(t)$ is an $n-$ dimensional vector, and

$$\mathbf{f}(t) = \boldsymbol{\zeta}(t) \quad \text{for} \quad t \leq t_0 \qquad (2.19)$$

The delayed vector $\mathbf{f}(t - \tau)$ over $t \in [t_0, t_f]$ may be approximated in terms of BPFs as

$$\mathbf{f}(t - \tau) \simeq \sum_{i=0}^{m-1} \mathbf{f}_i^\star(\tau) B_i(t) = F^\star(\tau)\mathbf{B}(t) \qquad (2.20)$$

where

$$\mathbf{f}_i^\star(\tau) = \frac{1}{T} \int_{t_0+iT}^{t_0+(i+1)T} \mathbf{f}(t - \tau)dt = \left\{ \begin{array}{ll} \boldsymbol{\zeta}_i & \text{for} \quad i < \mu \\ \mathbf{f}_{i-\mu} & \text{for} \quad i \geq \mu \end{array} \right. \qquad (2.21)$$

$$\boldsymbol{\zeta}_i = \frac{1}{T} \int_{t_0+iT}^{t_0+(i+1)T} \boldsymbol{\zeta}(t - \tau)dt \qquad \text{for} \quad i < \mu, \qquad (2.22)$$

μ is the number of BPFs considered over $t_0 \leq t \leq t_0 + \tau$, and

$$F^\star(\tau) = \left[\begin{array}{cccc} \mathbf{f}_0^\star(\tau), & \mathbf{f}_1^\star(\tau), & \cdots, & \mathbf{f}_{m-1}^\star(\tau) \end{array} \right] \qquad (2.23)$$

Let

$$\mathbf{f}(t) \simeq \sum_{i=0}^{m-1} \mathbf{f}_i B_i(t) = F\mathbf{B}(t) \qquad (2.24)$$

with

$$F = \left[\, \mathbf{f}_0, \;\; \mathbf{f}_1, \;\; \ldots, \;\; \mathbf{f}_{m-1} \,\right] \qquad (2.25)$$

Then by letting

$$\hat{\mathbf{f}}^\star(\tau) = \begin{bmatrix} \mathbf{f}_0^\star(\tau) \\ \mathbf{f}_1^\star(\tau) \\ \vdots \\ \mathbf{f}_{m-1}^\star(\tau) \end{bmatrix} ; \quad \hat{\mathbf{f}} = \begin{bmatrix} \mathbf{f}_0 \\ \mathbf{f}_1 \\ \vdots \\ \mathbf{f}_{m-1} \end{bmatrix} ; \quad \text{and} \quad \hat{\boldsymbol{\zeta}} = \begin{bmatrix} \zeta_0 \\ \zeta_1 \\ \vdots \\ \zeta_{\mu-1} \end{bmatrix} \qquad (2.26)$$

$\hat{\mathbf{f}}^\star(\tau)$ can be expressed in terms of $\hat{\mathbf{f}}$ and $\hat{\boldsymbol{\zeta}}$ as follows:

$$\hat{\mathbf{f}}^\star = E(n, \mu)\hat{\boldsymbol{\zeta}} + D(n, \mu)\hat{\mathbf{f}} \qquad (2.27)$$

where E and D are called shift operational matrices [24], given by

$$E(n, \mu) = \begin{bmatrix} I_{n\mu \times n\mu} \\ \cdots\cdots\cdots \\ O_{nd \times n\mu} \end{bmatrix} ; \quad D(n, \mu) = \begin{bmatrix} O_{n\mu \times nd} & \vdots & O_{n\mu \times n\mu} \\ \cdots\cdots & \vdots & \cdots\cdots \\ I_{nd \times nd} & \vdots & O_{nd \times n\mu} \end{bmatrix} \qquad (2.28)$$

with $d = m - \mu$.

2.2.5 Representation of reverse time function vector in BPFs

A reverse time function vector $\mathbf{f}\,(t_0 + t_f - t)$ over $t \in [t_0, \, t_f]$ may be approximated in terms of BPFs as

$$\mathbf{f}\,(t_0 + t_f - t) \approx \sum_{i=0}^{m-1} \tilde{\mathbf{f}}_i B_i(t) = \tilde{F}\mathbf{B}(t) \qquad (2.29)$$

where

$$
\begin{aligned}
\tilde{\mathbf{f}}_i &= \frac{1}{T} \int_{t_0+iT}^{t_0+(i+1)T} \mathbf{f}(t_0 + t_f - t)\,dt \\
&= \frac{1}{T} \int_{t_0+(m-1-i)T}^{t_0+(m-i)T} \mathbf{f}(\tau)\,d\tau = \mathbf{f}_{m-1-i}
\end{aligned}
\tag{2.30}
$$

Hence

$$
\tilde{F} = F\tilde{I}
\tag{2.31}
$$

where

$$
\tilde{I} =
\begin{bmatrix}
0 & \cdots & 0 & 0 & 0 & 1 \\
0 & \cdots & 0 & 0 & 1 & 0 \\
0 & \cdots & 0 & 1 & 0 & 0 \\
0 & \cdots & 1 & 0 & 0 & 0 \\
\vdots & & \vdots & \vdots & \vdots & \vdots \\
1 & \cdots & 0 & 0 & 0 & 0
\end{bmatrix}
\tag{2.32}
$$

is an $m \times m$ matrix, called reverse time operational matrix [91] of BPFs.

2.3   Legendre Polynomials (LPs)

A set of LPs can be generated from the recurrence relation [59]

$$
P_{i+1}(x) = \frac{(2i+1)}{(i+1)}\, x\, P_i(x) - \frac{i}{(i+1)}\, P_{i-1}(x)
\tag{2.33}
$$

with

$$
P_0(x) = 1, \quad \text{and} \quad P_1(x) = x
\tag{2.34}
$$

for $i = 1, 2, 3, \ldots\ldots$

A function $f(x)$ that is square integrable on $x \in [-1,\, 1]$ can be expressed in terms of LPs as

$$
f(x) \approx \sum_{i=0}^{m-1} f_i P_i(x) = \mathbf{f}^T \mathbf{P}(x)
\tag{2.35}
$$

where

$$\mathbf{f} = \begin{bmatrix} f_0, & f_1, & \dots, & f_{m-1} \end{bmatrix}^T \tag{2.36}$$

is called Legendre spectrum of $f(x)$, and

$$\mathbf{P}(x) = \begin{bmatrix} P_0(x), & P_1(x), & \dots, & P_{m-1}(x) \end{bmatrix}^T \tag{2.37}$$

is called the LP vector. f_i in Eq. (2.35) is given by

$$f_i = \frac{(2i+1)}{2} \int_{-1}^{1} f(x) P_i(x) dx \tag{2.38}$$

2.4 Shifted Legendre Polynomials (SLPs)

Legendre polynomials are defined over the interval $[-1,\ 1]$ while the signals of interest are normally considered over an arbitrary time interval $[t_0,\ t_f]$ where t_0 and t_f may not necessarily be -1 and 1, respectively. To suit our need, the above LPs are redefined on the time interval $t \in [t_0,\ t_f]$ using the relationship $x = \frac{2(t-t_0)}{(t_f-t_0)} - 1$.

Now the shifted polynomials $P_i \left\{ \frac{2(t-t_0)}{(t_f-t_0)} - 1 \right\}$, denoted by $L_i(t)$, are called shifted Legendre polynomials (SLPs). Then the recurrence relation in Eq. (2.33) becomes [59]

$$L_{i+1}(t) = \frac{(2i+1)}{(i+1)} \varphi(t) L_i(t) - \frac{i}{(i+1)} L_{i-1}(t) \tag{2.39}$$

for $i = 1, 2, 3, \dots \dots$ with

$$\varphi(t) = \frac{2(t-t_0)}{(t_f-t_0)} - 1 \tag{2.40}$$

$$L_0(t) = 1, \quad \text{and} \quad L_1(t) = \varphi(t) \tag{2.41}$$

Note that

$$L_i(t_0) = (-1)^i \quad \text{and} \quad L_i(t_f) = 1 \tag{2.42}$$

for all i.

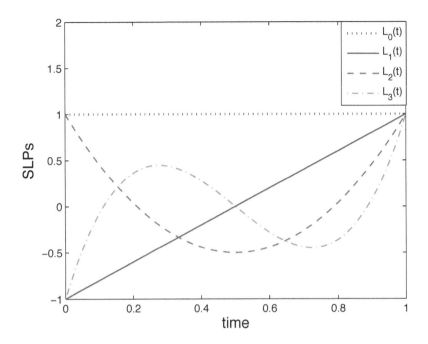

Figure 2.2: A set of four SLPs.

The first four SLPs over $t \in [0, 1]$ are shown in Fig. 2.2.

A function $f(t)$ that is square integrable on $t \in [t_0, t_f]$ can be represented in terms of SLPs as

$$f(t) \approx \sum_{i=0}^{m-1} f_i L_i(t) = \mathbf{f}^T \mathbf{L}(t) \qquad (2.43)$$

where

$$\mathbf{f} = \begin{bmatrix} f_0, & f_1, & \dots, & f_{m-1} \end{bmatrix}^T \qquad (2.44)$$

is called Legendre spectrum of $f(t)$, and

$$\mathbf{L}(t) = \begin{bmatrix} L_0(t), & L_1(t), & \dots, & L_{m-1}(t) \end{bmatrix}^T \qquad (2.45)$$

is called SLP vector. f_i in Eq. (2.43) is given by

$$f_i = \frac{(2i+1)}{(t_f - t_0)} \int_{t_0}^{t_f} f(t) L_i(t) dt \qquad (2.46)$$

2.4.1 Integration of $L(t)$

SLPs satisfy the relation

$$L_i(t) = \frac{(t_f - t_0)}{2(2i+1)} \left(\frac{d}{dt} L_{i+1}(t) - \frac{d}{dt} L_{i-1}(t) \right) \qquad (2.47)$$

for $i = 1, 2, 3, \ldots\ldots$.

Integrating $L_0(t)$ from t_0 to t, and expressing the result in terms of the same set of SLPs, we have

$$\int_{t_0}^{t} L_0(\tau)d\tau = \frac{(t_f - t_0)}{2} [L_0(t) + L_1(t)] \qquad (2.48)$$

Integrating Eq. (2.47) once with respect to t, and making use of the initial value of SLPs in Eq. (2.42) we obtain

$$\int_{t_0}^{t} L_i(\tau)d\tau = \frac{(t_f - t_0)}{2(2i+1)} [-L_{i-1}(t) + L_{i+1}(t)] \qquad (2.49)$$

Eqs. (2.48) and (2.49) can be written in the form

$$\int_{t_0}^{t} \mathbf{L}(\tau)d\tau \approx H_f \mathbf{L}(t) \qquad (2.50)$$

where

$$H_f = \frac{(t_f - t_0)}{2} \begin{bmatrix} 1 & 1 & 0 & 0 & \ldots & 0 & 0 \\ \frac{-1}{3} & 0 & \frac{1}{3} & 0 & \ldots & 0 & 0 \\ 0 & \frac{-1}{5} & 0 & \frac{1}{5} & \ldots & 0 & 0 \\ \vdots & \vdots & \vdots & \vdots & & \vdots & \vdots \\ 0 & 0 & 0 & 0 & \ldots & 0 & \frac{1}{2m-3} \\ 0 & 0 & 0 & 0 & \ldots & \frac{-1}{2m-1} & 0 \end{bmatrix} \qquad (2.51)$$

which is called the operational matrix of forward integration [26] of SLPs. Similarly if we integrate $\mathbf{L}(t)$ from t_f to t, make use of final value of SLPs in Eq. (2.42) and express the result in terms of the same set of SLPs, we have

$$\int_{t_f}^{t} \mathbf{L}(\tau)d\tau \approx H_b \mathbf{L}(t) \qquad (2.52)$$

where

$$
H_b = \frac{(t_f - t_0)}{2}
\begin{bmatrix}
-1 & 1 & 0 & 0 & \cdots & 0 & 0 \\
\frac{-1}{3} & 0 & \frac{1}{3} & 0 & \cdots & 0 & 0 \\
0 & \frac{-1}{5} & 0 & \frac{1}{5} & \cdots & 0 & 0 \\
\vdots & \vdots & \vdots & \vdots & & \vdots & \vdots \\
0 & 0 & 0 & 0 & \cdots & 0 & \frac{1}{2m-3} \\
0 & 0 & 0 & 0 & \cdots & \frac{-1}{2m-1} & 0
\end{bmatrix}
\tag{2.53}
$$

which is called the operational matrix of backward integration [28] of SLPs.

2.4.2 Product of two SLPs

The product of two SLPs $L_i(t)$ and $L_j(t)$ can be expressed [31] as

$$
L_i(t)L_j(t) \simeq \sum_{k=0}^{m-1} \psi_{ijk} L_k(t),
\tag{2.54}
$$

where

$$
\psi_{ijk} = \frac{(2k+1)}{(t_f - t_0)} \int_{t_0}^{t_f} L_i(t)L_j(t)L_k(t)dt.
\tag{2.55}
$$

Let

$$
\pi_{ijk} = \int_{t_0}^{t_f} L_i(t)L_j(t)L_k(t)dt.
\tag{2.56}
$$

Then

$$
\psi_{ijk} = \frac{(2k+1)}{(t_f - t_0)}\pi_{ijk}.
\tag{2.57}
$$

Notice that

$$
\pi_{ijk} = \pi_{ikj} = \pi_{jik} = \pi_{jki} = \pi_{kji} = \pi_{kij}
\tag{2.58}
$$

Also

$$
L_i(t)L_j(t) = \sum_{l=0}^{j} \frac{a_l a_{j-l} a_{i-j+l}}{a_{i+l}} \frac{2(i-j+2l)+1}{2(i+l)+1} L_{i-j+2l}(t)
\tag{2.59}
$$

where $i \geq j$, and

$$a_0 = 1, \quad a_{l+1} = \frac{(2l+1)}{(l+1)} a_l, \quad \text{for} \quad l = 0, 1, 2, \ldots \ldots \quad (2.60)$$

Moreover, for $i \geq j$

$$\pi_{ijk} = \left\{ \begin{array}{cc} \frac{a_l a_{j-l} a_{i-j+l}}{a_{i+l}} \frac{(t_f - t_0)}{2(i+l)+1} & \text{if} \quad k = i - j + 2l \\ 0 & \text{if} \quad k \neq i - j + 2l \end{array} \right\}. \quad (2.61)$$

2.4.3 Representation of $C(t)\mathbf{f}(t)$ in terms of SLPs

Assuming that $C(t)$ is square-integrable over $t \in [t_0, \ t_f]$, it can be expressed in terms of SLPs as

$$C(t) \simeq \sum_{i=0}^{m-1} C_i L_i(t). \quad (2.62)$$

Then

$$C(t)\mathbf{f}(t) = \mathbf{g}(t) \simeq \sum_{i=0}^{m-1} \mathbf{g}_i L_i(t) = G\mathbf{L}(t). \quad (2.63)$$

Also,

$$C(t)\mathbf{f}(t) \simeq \sum_{j=0}^{m-1} C_j L_j(t) \sum_{k=0}^{m-1} \mathbf{f}_k L_k(t). \quad (2.64)$$

Therefore [31]

$$\begin{aligned} \mathbf{g}_i &= \frac{(2i+1)}{(t_f - t_0)} \int_{t_0}^{t_f} \sum_{j=0}^{m-1} \sum_{k=0}^{m-1} C_j \mathbf{f}_k L_i(t) L_j(t) L_k(t) dt \\ &= \frac{(2i+1)}{(t_f - t_0)} \sum_{j=0}^{m-1} \sum_{k=0}^{m-1} \pi_{ijk} C_j \mathbf{f}_k. \end{aligned} \quad (2.65)$$

2.4.4 Representation of a time-delay vector function in SLPs

The SLP representation of $\mathbf{f}(t - \tau)$ is given by

$$\mathbf{f}(t - \tau) \simeq \sum_{i=0}^{m-1} \mathbf{f}_i^\star(\tau) L_i(t) = F^\star(\tau) \mathbf{L}(t) \qquad (2.66)$$

where

$$\begin{aligned} \mathbf{f}_i^\star(\tau) &= \frac{(2i+1)}{(t_f - t_0)} \int_{t_0}^{t_f} \mathbf{f}(t - \tau) L_i(t) dt \\ &= \boldsymbol{\zeta}_i(\tau) + \frac{(2i+1)}{(t_f - t_0)} \int_{t_0}^{t_f - \tau} \mathbf{f}(t) L_i(t + \tau) dt \quad (2.67) \end{aligned}$$

with

$$\boldsymbol{\zeta}_i(\tau) = \frac{(2i+1)}{(t_f - t_0)} \int_{t_0}^{t_0 + \tau} \boldsymbol{\zeta}(t - \tau) L_i(t) dt \qquad (2.68)$$

Now we express $L_i(t + \tau)$ in terms of SLPs as

$$L_i(t + \tau) = \sum_{j=0}^{i} \lambda_{ij}(\tau) L_j(t) = \boldsymbol{\lambda}_i^T(\tau) \mathbf{L}(t) \qquad (2.69)$$

with

$$\lambda_{ij}(\tau) = 0 \quad \text{for} \quad j > i. \qquad (2.70)$$

Now

$$\mathbf{f}(t) \approx \sum_{i=0}^{m-1} \mathbf{f}_i L_i(t) = F \mathbf{L}(t) = \left(\mathbf{L}^T(t) \otimes I_n \right) \hat{\mathbf{f}} \qquad (2.71)$$

where $\otimes$ is the Kronecker product [12] of matrices. Substituting Eqs. (2.69) and (2.71) into Eq. (2.67), we have

$$\begin{aligned} \mathbf{f}_i^\star(\tau) &= \boldsymbol{\zeta}_i(\tau) + \frac{(2i+1)}{(t_f - t_0)} \boldsymbol{\lambda}_i^T(\tau) \int_{t_0}^{t_f - \tau} \mathbf{L}(t) \left(\mathbf{L}^T(t) \otimes I_n \right) dt\, \hat{\mathbf{f}} \\ &= \boldsymbol{\zeta}_i(\tau) + \frac{(2i+1)}{(t_f - t_0)} \boldsymbol{\lambda}_i^T(\tau) \int_{t_0}^{t_f - \tau} \left(\mathbf{L}(t) \mathbf{L}^T(t) \otimes I_n \right) dt\, \hat{\mathbf{f}} \end{aligned}$$

$$(2.72)$$

Let

$$\Delta \;=\; \mathrm{diag}\left[\, 1,\; 3,\; \ldots,\; (2m-1)\,\right]/(t_f - t_0) \tag{2.73}$$

$$\Lambda(\tau) \;=\; \begin{bmatrix} \lambda_{00}(\tau) & 0 & 0 & \ldots & 0 \\ \lambda_{10}(\tau) & \lambda_{11}(\tau) & 0 & \ldots & 0 \\ \lambda_{20}(\tau) & \lambda_{21}(\tau) & \lambda_{22}(\tau) & \ldots & 0 \\ \vdots & \vdots & \vdots & & \vdots \\ \lambda_{m-1,0}(\tau) & \lambda_{m-1,1}(\tau) & \lambda_{m-1,2}(\tau) & \ldots & \lambda_{m-1,m-1}(\tau) \end{bmatrix} \tag{2.74}$$

be the time-advanced matrix of order $m \times m$, and

$$P(\tau) \;=\; \int_{t_0}^{t_f - \tau} \mathbf{L}(t)\mathbf{L}^T(t)\,dt. \tag{2.75}$$

Then Eq. (2.72) can be written in vector-matrix form as

$$\begin{aligned} \hat{\mathbf{f}}_i^\star(\tau) &= \hat{\boldsymbol{\zeta}}(\tau) + \Delta\,\Lambda(\tau)\left(P(\tau) \otimes I_n\right)\hat{\mathbf{f}} \\ &= \hat{\boldsymbol{\zeta}}(\tau) + \left(D(\tau) \otimes I_n\right)\hat{\mathbf{f}} \end{aligned} \tag{2.76}$$

where

$$D(\tau) \;=\; \Delta\,\Lambda(\tau)P(\tau) \tag{2.77}$$

is called the delay operational matrix [31] of SLPs, and

$$\hat{\boldsymbol{\zeta}}(\tau) \;=\; \begin{bmatrix} \boldsymbol{\zeta}_0(\tau) \\ \boldsymbol{\zeta}_1(\tau) \\ \vdots \\ \boldsymbol{\zeta}_{m-1}(\tau) \end{bmatrix} \tag{2.78}$$

To evaluate $D(\tau)$ we need to know $\Lambda(\tau)$ and $P(\tau)$.

2.4.5   Derivation of a time-advanced matrix of SLPs

The elements of a time-advanced matrix can be generated [31] in the following manner:

$$L_0(t + \tau) = 1 = L_0(t) \tag{2.79}$$

$$L_1(t + \tau) = \frac{2\tau}{(t_f - t_0)} + \frac{2(t - t_0)}{(t_f - t_0)} - 1 = \frac{2\tau}{(t_f - t_0)}L_0(t) + L_1(t) \tag{2.80}$$

$$\lambda_{00}(\tau) = 1 \tag{2.81}$$

$$\lambda_{10}(\tau) = \frac{2\tau}{(t_f - t_0)}; \quad \lambda_{11}(\tau) = 1 \tag{2.82}$$

$$(i + 1)L_{i+1}(t + \tau) = (2i + 1)L_1(t + \tau)L_i(t + \tau)$$
$$-iL_{i-1}(t + \tau) \tag{2.83}$$

$$(i + 1)\sum_{j=0}^{i+1}\lambda_{i+1,j}(\tau)L_j(t) = (2i + 1)\left[\frac{2\tau}{(t_f - t_0)} + L_1(t)\right]$$

$$\times \sum_{j=0}^{i}\lambda_{i,j}(\tau)L_j(t)$$

$$-i\sum_{j=0}^{i-1}\lambda_{i-1,j}(\tau)L_j(t) \tag{2.84}$$

$$L_1(t)L_j(t) = \frac{(j + 1)}{(2j + 1)}L_{j+1}(t) + \frac{j}{(2j + 1)}L_{j-1}(t) \tag{2.85}$$

$$(i+1)\sum_{j=0}^{i+1}\lambda_{i+1,j}(\tau)L_j(t) = \frac{2\tau(2i+1)}{(t_f-t_0)}\sum_{j=0}^{i}\lambda_{i,j}(\tau)L_j(t)$$

$$+(2i+1)\sum_{j=0}^{i}\lambda_{i,j}(\tau)\left[\frac{(j+1)}{(2j+1)}L_{j+1}(t) + \frac{j}{(2j+1)}L_{j-1}(t)\right]$$

$$-i\sum_{j=0}^{i-1}\lambda_{i-1,j}(\tau)L_j(t) \tag{2.86}$$

$$\lambda_{i+1,j}(\tau) = \frac{(2i+1)2\tau}{(i+1)(t_f-t_0)}\lambda_{i,j}(\tau) + \frac{j(2i+1)}{(i+1)(2j-1)}\lambda_{i,j-1}(\tau)$$

$$+\frac{(2i+1)(j+1)}{(i+1)(2j+3)}\lambda_{i,j+1}(\tau) - \frac{i}{(i+1)}\lambda_{i-1,j}(\tau) \tag{2.87}$$

where $\lambda_{i,j}(\tau) = 0$ for $i < 0$ or $j > i$

2.4.6 Algorithm for evaluating the integral in Eq. (2.75)

Let

$$p_{ij}(\tau) = \int_{t_0}^{t_f-\tau} L_i(t)L_j(t)dt \tag{2.88}$$

$$q_{ij}(\tau) = \int_{t_0}^{t_f-\tau} \dot{L}_i(t)L_j(t)dt \tag{2.89}$$

and

$$s_{ij}(\tau) = L_i(t_f-\tau)L_j(t_f-\tau) - (-1)^{i+j}. \tag{2.90}$$

Then

$$\frac{2(2i+1)}{(t_f-t_0)}p_{ij}(\tau) = -q_{i-1,j}(\tau) + q_{i+1,j}(\tau) \tag{2.91}$$

$$q_{i+1,j}(\tau) = q_{i-1,j}(\tau) + \frac{2(2i+1)}{(t_f-t_0)}p_{i,j}(\tau) \tag{2.92}$$

$$q_{ij}(\tau) = s_{ij}(\tau) - q_{ji}(\tau) \tag{2.93}$$

$$p_{0j}(\tau) = p_{j0}(\tau) = \begin{cases} \frac{(t_f-t_0)}{2}\left[L_0(t_f - \tau) + L_1(t_f - \tau)\right], & \text{for } j = 0 \\ \frac{(t_f-t_0)}{2(2j+1)}\left[-L_{j-1}(t_f - \tau) + L_{j+1}(t_f - \tau)\right], & (2.94) \\ \quad \text{for } j = 1, 2, 3 \dots \end{cases}$$

$$q_{0j}(\tau) = 0, \quad \text{for } j = 0, 1, 2 \dots\dots \qquad (2.95)$$

and

$$q_{1j}(\tau) = \frac{2}{(t_f - t_0)} p_{0j}(\tau). \qquad (2.96)$$

The algorithm [31] for evaluating the elements of matrix $P(\tau)$ is as follows:

step 1: Compute $p_{0j}(\tau)$ and $p_{j0}(\tau)$ for $j = 0, 1, \dots, 2(m - 1)$ using Eq. (2.94)

step 2: Set $q_{0j}(\tau) = 0$ for $j = 0, 1, \dots, 2(m - 1)$

step 3: Compute $q_{1j}(\tau)$ for $j = 0, 1, \dots, 2(m-1)$ using Eq. (2.96)

step 4: Compute $q_{i0}(\tau)$ and $q_{i1}(\tau)$ for $i = 0, 1, \dots, 2(m-1)$ using Eq. (2.93)

step 5: Set $i = 1$.

step 6: Compute $p_{ji}(\tau)$ and then $p_{ij}(\tau) = p_{ji}(\tau)$ for $j = i, i + 1, \dots, 2(m - 1) - i$ using Eq. (2.91)

step 7: If $i = m - 1$ then stop, else proceed.

step 8: Compute $q_{i+1,j}(\tau)$ for $j = i + 1, i + 2, \dots, 2(m - 1) - i$ using Eq. (2.92)

step 9: Compute $q_{j,i+1}(\tau)$ for $j = i + 2, i + 3, \dots, 2(m - 1) - i$ using Eq. (2.93)

step 10: Set $i = i + 1$ and go to step 6.

2.4.7 Representation of a reverse time function vector in SLPs

SLP representation of reverse time function vector $\mathbf{f}\,(t_0 + t_f - t)$ over $t \in [t_0, \ t_f]$ is given by

$$\mathbf{f}\,(t_0 + t_f - t) \approx \sum_{i=0}^{m-1} \tilde{\mathbf{f}}_i L_i(t) = \tilde{F}\mathbf{L}(t) \qquad (2.97)$$

where

$$
\begin{aligned}
\tilde{\mathbf{f}}_i &= \frac{(2i+1)}{(t_f - t_0)} \int_{t_0}^{t_f} \mathbf{f}(t_0 + t_f - t) L_i(t) dt \\
&= \frac{(2i+1)}{(t_f - t_0)} \int_{t_0}^{t_f} \mathbf{f}(\tau) L_i(t_0 + t_f - \tau) d\tau \\
&= \frac{(-1)^i (2i+1)}{(t_f - t_0)} \int_{t_0}^{t_f} \mathbf{f}(\tau) L_i(\tau) d\tau = (-1)^i \mathbf{f}_i \quad (2.98)
\end{aligned}
$$

Hence, we can write

$$\tilde{F} = F\tilde{I} \qquad (2.99)$$

with

$$
\tilde{I} =
\begin{bmatrix}
1 & 0 & 0 & 0 & \cdots & 0 \\
0 & -1 & 0 & 0 & \cdots & 0 \\
0 & 0 & 1 & 0 & \cdots & 0 \\
0 & 0 & 0 & -1 & \cdots & 0 \\
\vdots & \vdots & \vdots & \vdots & & \vdots \\
0 & 0 & 0 & 0 & \cdots & (-1)^{m-1}
\end{bmatrix}
\qquad (2.100)
$$

which is called the reverse time operational matrix [91] of SLPs.

2.5 Nonlinear Operational Matrix

$f\,(x(t),\,y(t))$ can be expressed in terms of LPs as

$$f\,(x(t),\,y(t)) \approx \sum_{i=0}^{n-1} P_i(x(t)) \sum_{j=0}^{n-1} f_{ij} P_j(y(t)) = \mathbf{P}^T(x(t)) F \mathbf{P}(y(t))$$

$$(2.101)$$

where F is the $n \times n$ Legendre spectrum of $f(x(t), y(t))$. Now we express $P_i(x(t))$ and $P_j(y(t))$ in terms of BPFs as

$$P_i(x(t)) \approx \sum_{k=0}^{m-1} n_{ik}(x)B_k(t) = \mathbf{n}_i^T(x)\mathbf{B}(t) \qquad (2.102)$$

$$P_j(y(t)) \approx \sum_{k=0}^{m-1} n_{jk}(y)B_k(t) = \mathbf{n}_j^T(y)\mathbf{B}(t) \qquad (2.103)$$

where $n_{ik}(x)$ and $n_{jk}(y)$ are the elements of $n \times m$ nonlinear operational matrices $N(x(t))$ and $N(y(t))$, respectively. Substituting Eqs. (2.102) and (2.103) into Eq. (2.101), we have

$$f(x(t), y(t)) \approx \sum_{i=0}^{n-1}\sum_{k=0}^{m-1} n_{ik}(x)B_k(t) \sum_{j=0}^{n-1} f_{ij} \sum_{l=0}^{m-1} n_{jl}(y)B_l(t)$$

$$\approx \sum_{i=0}^{n-1}\sum_{j=0}^{n-1}\sum_{k=0}^{m-1} n_{ik}(x)f_{ij}n_{jk}(y)B_k(t) \qquad (2.104)$$

after using the disjoint property of BPFs. From Eq. (2.34) we have

$$P_0(x(t)) = 1 = \mathbf{n}_0^T(x)\mathbf{B}(t) \qquad (2.105)$$

$$P_1(x(t)) = x(t) = \mathbf{n}_1^T(x)\mathbf{B}(t) \qquad (2.106)$$

where

$$\mathbf{n}_0(x) = \begin{bmatrix} 1, & 1, & \ldots, & 1 \end{bmatrix}^T, \quad \text{an } m-\text{ vector} \qquad (2.107)$$

$$\mathbf{n}_1(x) = \begin{bmatrix} x_0, & x_1, & \ldots, & x_{m-1} \end{bmatrix}^T \qquad (2.108)$$

Now Eq. (2.33) can be written as

$$\sum_{k=0}^{m-1} n_{i+1, k}(x)B_k(t) = \frac{(2i+1)}{(i+1)} \sum_{k=0}^{m-1} n_{1, k}(x)B_k(t) \sum_{l=0}^{m-1} n_{i, l}(x)B_l(t)$$

$$-\frac{i}{(i+1)} \sum_{k=0}^{m-1} n_{i-1, k}(x)B_k(t)$$

$$= \frac{(2i+1)}{(i+1)} \sum_{k=0}^{m-1} n_{1,k}(x)\, n_{i,k}(x) B_k(t) - \frac{i}{(i+1)} \sum_{k=0}^{m-1} n_{i-1,k}(x) B_k(t)$$

after using the disjoint property of BPFs. Hence

$$n_{i+1,k}(x) = \frac{(2i+1)}{(i+1)} n_{1,k}(x)\, n_{i,k}(x) - \frac{i}{(i+1)} n_{i-1,k}(x) \quad (2.109)$$

for $i = 1, 2, \ldots, n-2$ and $k = 0, 1, \ldots, m-1$.

It is now possible to compute the nonlinear operational matrix $N(x(t))$ recursively [95] using Eqs. (2.107)–(2.109).

2.6 Rationale for Choosing BPFs and SLPs

As already discussed, BPFs, HFs and WFs belong to the class of piecewise constant OFs. The rationale for choosing BPFs over HFs and WFs is as follows:

- HFs [22] and WFs [22] are to be generated, which is not the case with BPFs [22, 57] as they are all unity.

- The choice of m, the number of functions chosen over the time interval $[t_0, t_f]$, for HFs and WFs is 2^k, k being a positive integer, but for BPFs m can be any positive integer.

- Let the operational matrices of forward integration of HFs and WFs be denoted by P_H and P_W, and they are given [22] (for $m = 4$) by

$$P_H = \begin{bmatrix} \frac{1}{2} & -\frac{1}{4} & -\frac{1}{8\sqrt{2}} & -\frac{1}{8\sqrt{2}} \\ \frac{1}{4} & 0 & -\frac{1}{8\sqrt{2}} & \frac{1}{8\sqrt{2}} \\ \frac{1}{8\sqrt{2}} & \frac{1}{8\sqrt{2}} & 0 & 0 \\ \frac{1}{8\sqrt{2}} & -\frac{1}{8\sqrt{2}} & 0 & 0 \end{bmatrix}$$

$$P_W = \begin{bmatrix} \frac{1}{2} & -\frac{1}{4} & -\frac{1}{8} & 0 \\ \frac{1}{4} & 0 & 0 & -\frac{1}{8} \\ \frac{1}{8} & 0 & 0 & 0 \\ 0 & \frac{1}{8} & 0 & 0 \end{bmatrix}$$

By comparing the above matrices with the operational matrix of forward integration of BPFs given in Eq. (2.11), which is an upper triangular matrix, we observe that P_H and P_W have no such upper triangular structure. This structure is helpful in deriving recursive algorithms in the subsequent chapters.

In view of the above, BPFs are simpler and computationally more attractive than HFs and WFs.

The structure of forward integration operational matrix of GOPs is given [32] by

$$P = \frac{1}{p} \begin{bmatrix} B_0 - t_0 p - q & A_0 & 0 & 0 & \cdots & 0 & 0 & 0 \\ C_1 + D_1 & B_1 & A_1 & 0 & \cdots & 0 & 0 & 0 \\ D_2 & C_2 & B_2 & A_2 & \cdots & 0 & 0 & 0 \\ \vdots & \vdots & \vdots & \vdots & & \vdots & \vdots & \vdots \\ D_{m-2} & 0 & 0 & 0 & \cdots & C_{m-2} & B_{m-2} & A_{m-2} \\ D_{m-1} & 0 & 0 & 0 & \cdots & 0 & C_{m-1} & B_{m-1} \end{bmatrix}$$

where

$$p = \frac{2}{(t_f - t_0)} \quad \text{and} \quad q = -\frac{(t_f + t_0)}{(t_f - t_0)}$$

By selecting the appropriate values of the parameters A_r, B_r, C_r, D_r [32], the forward integration operational matrix P for each system of orthogonal polynomials can be obtained.

It is clear that the structure of forward integration operational matrix of SLPs is tridiagonal as the parameter D_r is zero, and hence it is more sparse than that of SCP1s, SCP2s, SJPs or HPs. So if we use the operational matrix of SLPs, it will be obviously more attractive computationally. Of course LaPs have a bidiagonal structure for its forward integration operational matrix. But LaPs and HPs are infinite range polynomials. So, to represent any signal accurately, the signal has to be known over infinite range, otherwise a large number of basis functions is required to attain a good accuracy.

Considering the application of orthogonal polynomials in the context of optimal control, the solutions provided by SCP1s, SCP2s, and SLPs [59] were found to be in better agreement with the exact solution than the other polynomials provide.

Moreover, the weighting function is not a constant, see Table 2.6 of [59], for all classes of orthogonal polynomials except SLPs for which it is unity.

It is shown [59] that SLPs, SCP2s and SCFs have a good filtering property. But, the convergence of Fourier series is not good enough at the points of discontinuity. A large number of SCFs is required to get a reasonable accuracy.

From the above discussion we say that SLPs and BPFs are the best among all classes of OFs. So SLPs and BPFs are employed in the study of state estimation and optimal control problems.

Chapter 3

State Estimation

Two recursive algorithms are presented for estimating state variables of observable linear time-invariant continuous-time dynamical systems from the input-output information using two classes of OFs, namely BPFs and SLPs. The principle of the Luenberger observer is utilized for estimating the state variables. The followed approach has the distinct advantage that the smoothing effect of integration reduces the influence of zero-mean observation noise on estimation. Results of a simulation study on two examples indicate that the recursive algorithms work well.

3.1 Introduction

State estimation plays an important role in the context of state feedback control as it requires complete and accurate information of all state variables. When this information is not available, it is estimated using the Luenberger observer [1]. In the case of noisy environment the Luenberger observer produces erroneous estimates of the state variables unless the measurement noise is filtered out.

Interestingly, the OF approach has an inherent filtering property [59] as it involves an integration process which has the smoothing effect. As it appears from the literature, two attempts have been made on the state estimation problem by using two different classes of OFs, i.e BPFs [23] and SCP1s [39] so far.

Upon examining critically the BPF approach [23] and SCP1 approach [39], it is observed that the BPF approach is purely recursive and uses multiple integration. The number of integrations increases as the order of the system increases, i.e for an nth order system the state equation has to be integrated n times, which is computationally not attractive. The recursive algorithm in [23] is restricted to observable canonical form as it directly deals with the elements of system matrices in this form.

Application of OFs has been extended to one more closely related problem in this decade. That is, costate estimation [64] has been studied via OFs. The basic idea of this continued research activity is to develop computationally elegant algorithms.

Let the structure of integration operational matrix P of SCP1s and SLPs be respectively denoted by P_C and P_L, and they are given [59] by

$$P_C = \begin{bmatrix} b_0 & a_1 & 0 & 0 & \cdots & 0 & 0 \\ b_1 & 0 & a_2 & 0 & \cdots & 0 & 0 \\ b_2 & c_1 & 0 & a_3 & \cdots & 0 & 0 \\ b_3 & 0 & c_2 & 0 & \cdots & 0 & 0 \\ \vdots & \vdots & \vdots & \vdots & & \vdots & \vdots \\ b_{m-2} & 0 & 0 & 0 & \cdots & 0 & a_{m-1} \\ b_{m-1} & 0 & 0 & 0 & \cdots & c_{m-2} & 0 \end{bmatrix} \tag{3.1}$$

and

$$
P_L =
\begin{bmatrix}
\bar{b}_0 & \bar{a}_1 & 0 & 0 & \cdots & 0 & 0 \\
\bar{c}_0 & 0 & \bar{a}_2 & 0 & \cdots & 0 & 0 \\
0 & \bar{c}_1 & 0 & \bar{a}_3 & \cdots & 0 & 0 \\
0 & 0 & \bar{c}_2 & 0 & \cdots & 0 & 0 \\
\vdots & \vdots & \vdots & \vdots & & \vdots & \vdots \\
0 & 0 & 0 & 0 & \cdots & 0 & \bar{a}_{m-1} \\
0 & 0 & 0 & 0 & \cdots & \bar{c}_{m-2} & 0
\end{bmatrix}
\tag{3.2}
$$

Observe that P_L is more sparse than P_C. So if we can somehow develop a recursive algorithm via SLPs, it will be obviously more elegant computationally.

Moreover, let a square integrable function $\bar{s}(t)$ on $t \in [t_0, \ t_f]$ be corrupted with a zero-mean observation noise $n(t)$ so that the measured signal is given by $\tilde{s}(t) = \bar{s}(t) + n(t)$. Then the ith Chebyshev coefficient in Chebyshev series representation of $\tilde{s}(t)$ is given by

$$
\tilde{s}_i = \alpha \int_{t_0}^{t_f} w(t)\phi_i(t)\tilde{s}(t)dt.
$$

where

$$
\alpha =
\begin{cases}
\frac{1}{\pi}, & i = 0, \\[2mm]
\frac{2}{\pi}, & i \neq 0,
\end{cases}
$$

$$
w(t) = \frac{1}{\sqrt{(t - t_0)(t_f - t)}}
$$

is the weighting function of SCP1s, and $\phi_i(t)$ is the ith shifted Chebyshev polynomial of first kind. Notice that $w(t) = \infty$ at $t = t_0$ and t_f. So $\tilde{s}_i$ can be evaluated using the open quadrature formula if $\tilde{s} = \bar{s}$, i.e. the signal, uncorrupted with noise in this case, is explicitly available. However, $\tilde{s}_i$ cannot be evaluated for $\tilde{s}(t)$ because $\tilde{s}(t)$ is not available explicitly in practice. This is the drawback of the SCP1 approach [59]. Therefore state estimation cannot be done via SCP1s in a noisy environment.

In view of the above observations, in this chapter we consider BPFs and SLPs, and present two recursive algorithms for state estimation. Compared to the recursive algorithms in [23, 39], the algorithms in [76, 78, 89] are computationally more elegant. The chapter is organised as follows. The next section deals with the inherent filtering property of OFs. Section 3.3 deals with state estimation via BPFs and SLPs. Two numerical examples are considered in Section 3.4 to demonstrate and compare the performance of recursive algorithms. Section 3.5 concludes the chapter.

3.2 Inherent Filtering Property of OFs

If the (block-pulse/Legendre) spectrum of a signal $\bar{s}(t)$ is $\{\bar{s}_k\}$, and if the signal corrupted with a zero mean random noise $n(t)$ is having $\{\tilde{s}_k\}$ as the (block-pulse/Legendre) spectrum of $\tilde{s}(t) = \bar{s}(t) + n(t)$, then $\tilde{s}_k \longrightarrow \bar{s}_k$ for all nonnegative integral values of k, if t is sufficiently large. In other words, if the (block-pulse/Legendre) spectrum of $\tilde{s}(t)$ is multiplied by the (BPF/SLP) basis vector, the original signal $\bar{s}(t)$ may be recovered from the noisy signal.

Fig. 3.1 shows the schematic diagram of a signal reconstructor which is simply software loaded in a digital computer and it computes the spectrum of corrupted signal $\tilde{s}(t)$ with respect to the chosen orthogonal system (BPFs/SLPs) and multiplies the spectrum by the corresponding basis (BPF/SLP) vector to produce $\hat{s}(t)$, the reconstructed signal. This $\hat{s}(t)$ is expected to follow $\bar{s}(t)$. The function $e(t) = \bar{s}(t) - \hat{s}(t)$ in Fig. 3.1 represents the residual error.

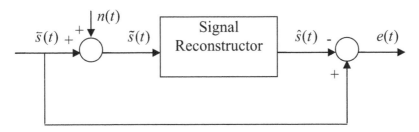

Figure 3.1: Schematic diagram to study the inherent filtering property of OFs.

Let us now demonstrate the inherent filtering property of SLPs by considering the signal

$$\bar{s}(t) = -0.3143 \cos(0.98t + 0.3399) + 0.2227 \cos(0.23t + 1.2302)$$
$$-0.2758 \cos(2.76t - 0.5961) \tag{3.3}$$

over $t \in [0, 1]$, $n(t)$ as a zero mean Gaussian noise with noise-to-signal ratio (NSR) equal to $0.0, 0.1, 0.2, 0.3, 0.4$, or 0.5, and the first eight SLPs.

Fig. 3.2 shows the signal $\bar{s}(t)$, the corrupted signal $\tilde{s}(t)$ with NSR $= 0.1$, the reconstructed signal $\hat{s}(t)$, and the residual error $e(t)$. Fig. 3.3 shows $e(t)$ at different noise levels. It is observed that the maximum absolute value (≈ 0.025) of $e(t)$, for NSR $= 0.5$, is approximately five percent of maximum absolute value (≈ 0.5) of $\bar{s}(t)$. This demonstrates clearly the inherent filtering property of SLPs.

3.3 State Estimation

We consider an observable system of order n.

$$\dot{\mathbf{x}}(t) = A\mathbf{x}(t) + B\mathbf{u}(t), \quad \mathbf{x}(t_0) = \tilde{\mathbf{x}}_0 \tag{3.4}$$
$$\mathbf{y}(t) = C\mathbf{x}(t) \tag{3.5}$$

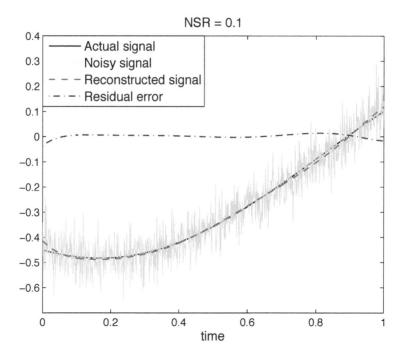

Figure 3.2: Filtering property of SLPs.

where $\mathbf{u}(t)$ is the r−dimensional control vector, $\mathbf{y}(t)$ the p−dimensional output vector with $p < n$, and A, B and C are matrices of appropriate dimensions. The objective is to estimate the state vector $\mathbf{x}(t)$ from the knowledge of $\mathbf{u}(t)$ and $\mathbf{y}(t)$ using the Luenberger observer [1].

The state equations of the Luenberger observer are given by

$$
\begin{aligned}
\dot{\hat{\mathbf{x}}}(t) &= A\hat{\mathbf{x}}(t) + B\mathbf{u}(t) + K[\mathbf{y}(t) - C\hat{\mathbf{x}}(t)] \\
&= \tilde{A}\hat{\mathbf{x}}(t) + B\mathbf{u}(t) + K\mathbf{y}(t)
\end{aligned} \tag{3.6}
$$

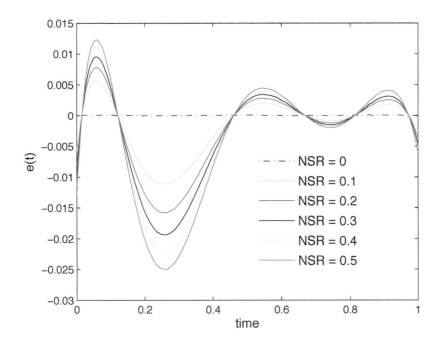

Figure 3.3: Residual error in signal reconstruction by SLPs.

where

$$\tilde{A} = A - KC \tag{3.7}$$

and K is the observer gain matrix which is selected in such a manner that the eigenvalues of the observer, given by the roots of the characteristic polynomial

$$r(s) = |sI - A + KC| = s^n + \tilde{a}_{n-1} s^{n-1} + \ldots + \tilde{a}_1 s + \tilde{a}_0 \tag{3.8}$$

lie at chosen locations in the left half of $s-$plane. $\hat{\mathbf{x}}(t)$ is the estimated state vector of order n.

Expressing $\hat{\mathbf{x}}(t)$, $\mathbf{u}(t)$, $\mathbf{y}(t)$ and $\hat{\mathbf{x}}(t_0)$ in terms of $m-$set of OFs, either BPF vector ($\mathbf{B}(t)$) or SLP vector ($\mathbf{L}(t)$), we have

$$\hat{\mathbf{x}}(t) \approx \sum_{i=0}^{m-1} \hat{\mathbf{x}}_i \phi_i(t) = \hat{X}\boldsymbol{\phi}(t) \tag{3.9}$$

$$\mathbf{u}(t) \approx \sum_{i=0}^{m-1} \mathbf{u}_i \phi_i(t) = U\boldsymbol{\phi}(t) \tag{3.10}$$

$$\mathbf{y}(t) \approx \sum_{i=0}^{m-1} \mathbf{y}_i \phi_i(t) = Y\boldsymbol{\phi}(t) \tag{3.11}$$

$$\hat{\mathbf{x}}(t_0) \approx \hat{X}_0\boldsymbol{\phi}(t) \tag{3.12}$$

where

$$\hat{X} = \begin{bmatrix} \hat{\mathbf{x}}_0, & \hat{\mathbf{x}}_1, & \dots, & \hat{\mathbf{x}}_{m-1} \end{bmatrix} \tag{3.13}$$

$$U = \begin{bmatrix} \mathbf{u}_0, & \mathbf{u}_1, & \dots, & \mathbf{u}_{m-1} \end{bmatrix} \tag{3.14}$$

$$Y = \begin{bmatrix} \mathbf{y}_0, & \mathbf{y}_1, & \dots, & \mathbf{y}_{m-1} \end{bmatrix} \tag{3.15}$$

$$\hat{X}_0 = \begin{bmatrix} \hat{\mathbf{x}}(t_0), & \hat{\mathbf{x}}(t_0), & \dots, & \hat{\mathbf{x}}(t_0) \end{bmatrix} \quad \text{if} \quad \boldsymbol{\phi}(t) = \mathbf{B}(t) \tag{3.16}$$

$$= \begin{bmatrix} \hat{\mathbf{x}}(t_0), & \mathbf{0}, & \mathbf{0}, & \dots, & \mathbf{0} \end{bmatrix} \quad \text{if} \quad \boldsymbol{\phi}(t) = \mathbf{L}(t) \tag{3.17}$$

Integrating Eq. (3.6) once with respect to t, we obtain

$$\hat{\mathbf{x}}(t) - \hat{\mathbf{x}}(t_0) = \int_{t_0}^{t} \left[\tilde{A}\hat{\mathbf{x}}(\tau) + B\mathbf{u}(\tau) + K\mathbf{y}(\tau) \right] d\tau \tag{3.18}$$

Substituting Eqs. (3.9)–(3.12) into Eq. (3.18), applying Eq. (2.10)/(2.50), and simplifying, lead to

$$\hat{X} = \left(\tilde{A}\hat{X} + BU + KY \right) H_f + \hat{X}_0 = \tilde{A}\hat{X}H_f + V \tag{3.19}$$

where

$$V = (BU + KY) H_f + \hat{X}_0 = \begin{bmatrix} \mathbf{v}_0, & \mathbf{v}_1, & \dots, & \mathbf{v}_{m-1} \end{bmatrix} \tag{3.20}$$

Eq. (3.19) is to be solved for $\hat{X}$. Obtaining solution $\hat{X}$ is discussed in the following subsections.

3.3.1 Kronecker product method

Eq. (3.19) can be written as

$$\text{Vec}\left(\hat{X}\right) = \left(I_{mn} - H_f^T \otimes \tilde{A}\right)^{-1}\text{Vec}(V) \qquad (3.21)$$

where $\text{Vec}(\cdot)$ is a vector valued function [12] of a matrix, and I_{mn} is the identity matrix of order mn, and $\otimes$ is the Kronecker product [12] of matrices. Though this method [76, 78, 89] is straightforward, it is computationally not elegant as it involves inversion of a matrix of size mn which becomes large for the large value of m. Note that the accuracy of end result depends on the m value.

3.3.2 Recursive algorithm via BPFs

Substitute matrix H_f of BPFs, see Eq. (2.11), into Eq. (3.19) and simplify to obtain the following recursive algorithm [78, 89]:

$$\hat{\mathbf{x}}_0 = \left(I_n - 0.5T\tilde{A}\right)^{-1}\mathbf{v}_0 \qquad (3.22)$$

$$\hat{\mathbf{x}}_i = \left(I_n - 0.5T\tilde{A}\right)^{-1}\left[\left(I_n + 0.5T\tilde{A}\right)\hat{\mathbf{x}}_{i-1} + (\mathbf{v}_i - \mathbf{v}_{i-1})\right] \qquad (3.23)$$

for $i = 1, 2, 3, \ldots, m-1$,
where

$$\mathbf{v}_0 = 0.5T(B\mathbf{u}_0 + K\mathbf{y}_0) + \hat{\mathbf{x}}(t_0) \qquad (3.24)$$

$$\mathbf{v}_i - \mathbf{v}_{i-1} = 0.5T\left[B\left(\mathbf{u}_{i-1} + \mathbf{u}_i\right) + K\left(\mathbf{y}_{i-1} + \mathbf{y}_i\right)\right] \qquad (3.25)$$

for $i = 1, 2, 3, \ldots, m-1$.

Since $\hat{\mathbf{x}}(t_0)$ is unknown, it can be chosen arbitrarily and the estimation error will approach zero asymptotically at a rate determined by the chosen roots of the characteristic polynomial $r(s)$.

3.3.3 Recursive algorithm via SLPs

Substituting matrix H_f of SLPs (Eq. (2.51)) into Eq. (3.19) and rearranging the terms, we have

$$
\begin{bmatrix}
W_{00} & W_{01} & \bigcirc & \bigcirc & \cdots & \bigcirc & \bigcirc \\
W_{10} & W_{11} & W_{12} & \bigcirc & \cdots & \bigcirc & \bigcirc \\
\bigcirc & W_{21} & W_{22} & W_{23} & \cdots & \bigcirc & \bigcirc \\
\vdots & \vdots & \vdots & \vdots & & \vdots & \vdots \\
\bigcirc & \bigcirc & \bigcirc & \bigcirc & \cdots & W_{m-2,\,m-2} & W_{m-2,\,m-1} \\
\bigcirc & \bigcirc & \bigcirc & \bigcirc & \cdots & W_{m-1,\,m-2} & W_{m-1,\,m-1}
\end{bmatrix}
\begin{bmatrix}
\mathbf{s}_0 \\ \mathbf{s}_1 \\ \mathbf{s}_2 \\ \vdots \\ \mathbf{s}_{m-2} \\ \mathbf{s}_{m-1}
\end{bmatrix}
=
\begin{bmatrix}
\mathbf{v}_0 \\ \mathbf{v}_1 \\ \mathbf{v}_2 \\ \vdots \\ \mathbf{v}_{m-2} \\ \mathbf{v}_{m-1}
\end{bmatrix}
\tag{3.26}
$$

where

$$
W_{ij} =
\begin{cases}
\dfrac{2I_n}{(t_f - t_0)} - \tilde{A} & \text{if } i = j = 0 \\[2mm]
\dfrac{\tilde{A}}{(2i+3)} & \text{if } i = 0, 1, 2, \ldots, m-2 \text{ and } j = i+1 \\[2mm]
\dfrac{-\tilde{A}}{(2i-1)} & \text{if } i = 1, 2, 3, \ldots, m-1 \text{ and } j = i-1 \\[2mm]
\dfrac{2I_n}{(t_f - t_0)} & \text{if } i = j = 1, 2, \ldots, m-1 \\[2mm]
\bigcirc & \text{otherwise}
\end{cases}
\tag{3.27}
$$

$$
\mathbf{s}_i = \hat{\mathbf{x}}_i \quad \text{for} \quad i = 0, 1, 2, \ldots, m-1
\tag{3.28}
$$

and

$$
\mathbf{v}_i =
\begin{cases}
\dfrac{2}{(t_f-t_0)}\hat{\mathbf{x}}(t_0) + (B\mathbf{u}_0 + K\mathbf{y}_0) - \\
\quad \frac{1}{3}(B\mathbf{u}_1 + K\mathbf{y}_1) & \text{if } i = 0 \\[2mm]
\dfrac{1}{(2i-1)}(B\mathbf{u}_{i-1} + K\mathbf{y}_{i-1}) - \\
\quad \dfrac{1}{(2i+3)}(B\mathbf{u}_{i+1} + K\mathbf{y}_{i+1}) & \text{if } i = 1, 2, \ldots, m-2 \\[2mm]
\dfrac{1}{(2m-3)}(B\mathbf{u}_{m-2} + K\mathbf{y}_{m-2}) & \text{if } i = m-1
\end{cases}
\tag{3.29}
$$

The following recursive algorithm [76, 89] follows from Eqs. (3.26)–(3.29):

$$
\mathbf{d}_i = \begin{cases} M_{ii}\mathbf{v}_i & \text{if } i = m - 1 \\ M_{ii}\left(\mathbf{v}_i - W_{i,i+1}\mathbf{d}_{i+1}\right) & \text{if } i = m - 2, m - 3, \ldots, 1, 0. \end{cases} \tag{3.30}
$$

$$
R_{i,i-1} = -M_{ii}W_{i,i-1} \qquad \text{if } i = m - 1, m - 2, \ldots, 2, 1. \tag{3.31}
$$

$$
M_{ii} = \begin{cases} W_{ii}^{-1} & \text{if } i = m - 1 \\ \left(W_{ii} + W_{i,i+1}R_{i+1,i}\right)^{-1} & \text{if } i = m - 2, m - 3, \ldots, 1, 0. \end{cases} \tag{3.32}
$$

$$
\mathbf{s}_0 = \mathbf{d}_0 \tag{3.33}
$$

$$
\mathbf{s}_i = R_{i,i-1}\mathbf{s}_{i-1} + \mathbf{d}_i \qquad \text{if } i = 1, 2, \ldots, m - 1. \tag{3.34}
$$

3.3.4 Modification of the recursive algorithm of Sinha and Qi-Jie

Sinha and Qi-Jie [23] assumed $\hat{\mathbf{x}}(t_0) = \mathbf{0}$ and developed the recursive algorithm. In order to deal with the nonzero initial state vector, here we modify the algorithm in [23]. They consider dynamical equations of a single-input single-output system of order n in observable canonical form, given by

$$
\begin{bmatrix} \dot{x}_1(t) \\ \dot{x}_2(t) \\ \dot{x}_3(t) \\ \vdots \\ \dot{x}_n(t) \end{bmatrix} = \begin{bmatrix} 0 & 0 & \cdots & 0 & a_0 \\ 1 & 0 & \cdots & 0 & a_1 \\ 0 & 1 & \cdots & 0 & a_2 \\ \vdots & \vdots & & \vdots & \vdots \\ 0 & 0 & \cdots & 1 & a_{n-1} \end{bmatrix} \begin{bmatrix} x_1(t) \\ x_2(t) \\ x_3(t) \\ \vdots \\ x_n(t) \end{bmatrix} + \begin{bmatrix} b_0(t) \\ b_1(t) \\ b_2(t) \\ \vdots \\ b_{n-1}(t) \end{bmatrix} u(t)
$$

$$
\tag{3.35}
$$

$$
y(t) = \begin{bmatrix} 0 & 0 & \cdots & 0 & 1 \end{bmatrix} \begin{bmatrix} x_1(t) & x_2(t) & \cdots & x_{n-1}(t) & x_n(t) \end{bmatrix}^T
$$

$$
\tag{3.36}
$$

The recursive algorithm obtained via BPFs is as follows:

$$D_n \hat{x}_{n,i} = \sum_{j=1}^{n} \tilde{a}_{n-j} \left[I_{j,i-1}(\hat{x}_n) + \sum_{l=0}^{n-2} T(0.5T)^l I_{j-l-1,i-1}(\hat{x}_n) \right]$$

$$+ \sum_{j=1}^{n} \tilde{a}_{n-j} (0.5T)^j \hat{x}_{n,i-1}$$

$$+ \sum_{j=1}^{n} [k_{j-1} I_{n+1-j,i}(y) + b_{j-1} I_{n+1-j,i}(u)]$$

$$+ \sum_{j=1}^{n-1} I_{n-j,i} [x_j(0)] + x_n(0) \tag{3.37}$$

where

$$\tilde{a}_n = a_n - k_n \tag{3.38}$$

$$D_n = 1 - \sum_{j=1}^{n} \tilde{a}_{n-j} (0.5T)^j \tag{3.39}$$

and $I_{n,i}(\hat{x}_n)$ is the ith term in the BPF expansion of the nth integral of $\hat{x}_n$.

$$I_{j-l-1,i-1}(\hat{x}_n) = 0 \text{ for } j-l-1 \le 0, \, i-1 < 0. \tag{3.40}$$

$$I_{l,i}(y) = I_{l,i-1}(y) + T I_{l-1,i-1}(y) + 0.5T^2 I_{l-2,i-1}(y)$$

$$+ 0.25 T^3 I_{l-3,i-1}(y) +$$

$$\cdots\cdots + \frac{T^{l-1}}{2^{l-2}} I_{1,i-1}(y) + (0.5T)^l (y_{i-1} + y_i) \tag{3.41}$$

Dividing Eq. (3.37) by D_n leads to the estimation of the last state. The other states can be estimated in terms of the last state through the following equation:

$$
\hat{x}_{j,i} = \sum_{l=1}^{j} [\tilde{a}_{l-1} I_{j-l+1,i} (\hat{x}_n) + k_{l-1} I_{j-l+1,i}(y) + b_{l-1} I_{j-l+1,i}(u)]
$$
$$
+ \sum_{l=1}^{j-1} I_{j-l,i} [x_l(0)] + x_j(0) \tag{3.42}
$$

for $j = 1, 2, \ldots, n-1$.

3.4   Illustrative Examples

Two examples are considered here to illustrate the modified recursive algorithm, and the recursive algorithms derived via BPFs and SLPs.

Example 3.1

Consider the second-order system [23]

$$
\begin{bmatrix} \dot{x}_1(t) \\ \dot{x}_2(t) \end{bmatrix} = \begin{bmatrix} 0 & -2 \\ 1 & -3 \end{bmatrix} \begin{bmatrix} x_1(t) \\ x_2(t) \end{bmatrix} + \begin{bmatrix} 0 \\ 1 \end{bmatrix} u(t) \tag{3.43}
$$

$$
y(t) = \begin{bmatrix} 0 & 1 \end{bmatrix} \begin{bmatrix} x_1(t) & x_2(t) \end{bmatrix}^T \tag{3.44}
$$

$$
u(t) = 2\cos 0.23t - \cos 0.98t - \cos 2.76t \tag{3.45}
$$

The observer is designed to have both the eigenvalues at $s = -6$ by choosing $k_1 = 34$ and $k_2 = 9$. Upon solving, the state variables are given by

$$
\begin{aligned}
x_1(t) &= \left[\, 2x_1(0) - 2x_2(0) + 2.5467484 \,\right] e^{-t} + \left[\, -x_1(0) + 2x_2(0) \right. \\
&\quad \left. -0.8232038 \,\right] e^{-2t} - 1.9363483 \cos\left(0.23t - 0.3405653\right) \\
&\quad + 0.641356 \cos\left(0.98t - 1.2309133\right) \\
&\quad - 0.1998842 \cos\left(2.76t + 0.9746775\right) \quad\quad\quad\quad (3.46) \\
x_2(t) &= \left[\, x_1(0) - x_2(0) + 1.2733742 \,\right] e^{-t} + \left[\, -x_1(0) + 2x_2(0) \right. \\
&\quad \left. -0.8232036 \,\right] e^{-2t} + 0.22268 \cos\left(0.23t + 1.2302309\right) \\
&\quad - 0.3142645 \cos\left(0.98t + 0.339883\right) \\
&\quad - 0.2758402 \cos\left(2.76t - 0.5961185\right) \quad\quad\quad\quad (3.47)
\end{aligned}
$$

The expressions of $x_1(t)$ and $x_2(t)$ are incorrect in [23]. Consequently, the results shown in Table 1 of [23] are incorrect.

First, assuming that

$$
\hat{\mathbf{x}}(t_0) = \mathbf{x}(0) = \left[\, -1.7235446 \quad -0.4501704 \,\right]^T
$$

the state estimates $\hat{x}_1(t)$ and $\hat{x}_2(t)$ are found by using the the modified recursive algorithm and the recursive algorithm with $m = 8$ and sampling interval $\Delta T = 0.005$ over the interval $t \in [0, 1]$. The estimates are shown in Figs. 3.4 and 3.5.

Next, since the initial conditions $x_1(0)$ and $x_2(0)$ are unknown in general, for the sake of simplicity they are assumed to be $x_1(0) = -1.7235446$ and $x_2(0) = -0.4501704$, so that the exponential terms in Eqs. (3.46) and (3.47) vanish.

Assuming that $\hat{x}_1(0) = \hat{x}_2(0) = 0$, the state estimates $\hat{x}_1(t)$ and $\hat{x}_2(t)$ are found by the recursive algorithm and the the modified recursive algorithm, and shown in Figs. 3.6 and 3.7. As can be seen from Figs. 3.6 and 3.7, the estimates obtained via BPF algorithms (Section 3.3.2 and Section 3.3.4) are just identical as they

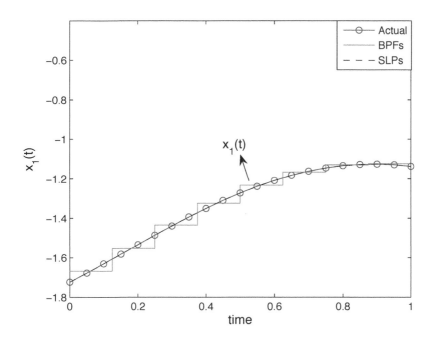

Figure 3.4: Estimate of state-1 with known initial states.

are superimposed. In both the cases, the estimates obtained are piecewise constant which is the inherent characteristic of BPFs. The state estimates obtained via SLPs are smooth and are shown in Figs. 3.6 and 3.7. In general, the computational time and work-space memory are relatively higher compared to that of BPFs as SLPs are needed to be generated and used to compute $\hat{\mathbf{x}}(t)$, $\mathbf{u}(t)$, and $\mathbf{y}(t)$ using Eqs. (3.9)−(3.11). Since BPFs are all unity, they need not be specifically computed, and this leads to significant saving in computational time and work-space memory in the case of BPF approach.

In addition to the above noise-free case, system output corrupted with measurement noise (having Gaussian distribution with zero-mean) is considered with NSR equal to 0.1, and the

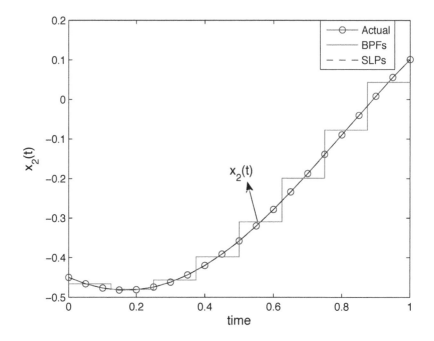

Figure 3.5: Estimate of state-2 with known initial states.

states are estimated again to obtain the estimates as shown in Figs. 3.8 and 3.9. The estimates in this case are found to be slightly different from the ones obtained in the noise-free case.

Example 3.2

Consider the third-order system [23]

$$\begin{bmatrix} \dot{x}_1(t) \\ \dot{x}_2(t) \\ \dot{x}_3(t) \end{bmatrix} = \begin{bmatrix} 0 & 0 & -6 \\ 1 & 0 & -11 \\ 0 & 1 & -6 \end{bmatrix} \begin{bmatrix} x_1(t) \\ x_2(t) \\ x_3(t) \end{bmatrix} + \begin{bmatrix} 1 \\ 0 \\ 0 \end{bmatrix} u(t) \quad (3.48)$$

$$y(t) = \begin{bmatrix} 0 & 0 & 1 \end{bmatrix} \begin{bmatrix} x_1(t) & x_2(t) & x_3(t) \end{bmatrix}^T \quad (3.49)$$

The same input $u(t)$ in Eq. (3.45) is used here also. The observer is designed to have eigenvalues at $s = -6$ by choosing the observer gain elements $k_1 = 210$, $k_2 = 97$ and $k_3 = 12$.

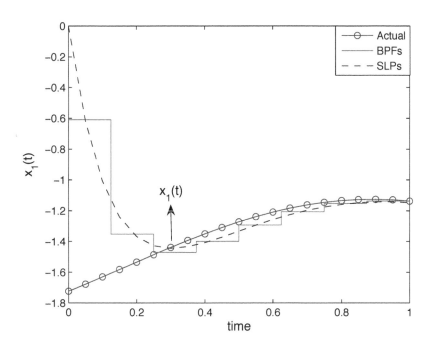

Figure 3.6: Estimate of state-1.

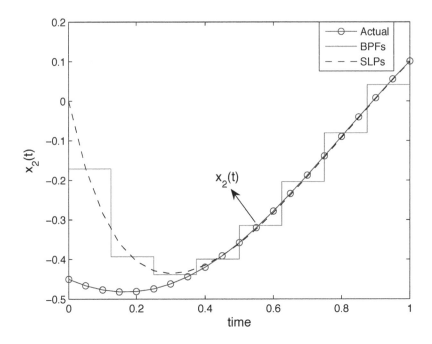

Figure 3.7: Estimate of state-2.

The actual state variables are found to be

$$
\begin{aligned}
x_1(t) &= \left[\, 3x_1(0) - 3x_2(0) + 3x_3(0) - 3.8201222\,\right] e^{-t} + \left[\, -3x_1(0)\right. \\
&\quad \left. + 6x_2(0) - 12x_3(0) + 1.2348053\,\right] e^{-2t} + \left[\, x_1(0) - 3x_2(0)\right. \\
&\quad \left. + 9x_3(0) - 0.1810469\,\right] e^{-3t} + 3.5504412\cos\left(\, 0.23t\right. \\
&\quad \left. - 0.291683\,\right) - 1.1821942\cos\left(\, 0.98t - 1.0168277\,\right) \\
&\quad - 0.414381\cos\left(\, 2.76t - 1.5413544\,\right)
\end{aligned}
\tag{3.50}
$$

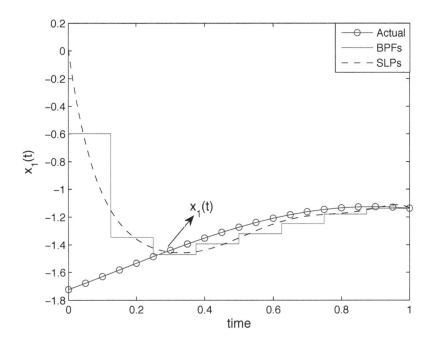

Figure 3.8: Estimate of state-1 with NSR $= 0.1$.

$$
\begin{aligned}
x_2(t) = & \left[2.5x_1(0) - 2.5x_2(0) + 2.5x_3(0) - 3.1834352\right] e^{-t} \\
& + \left[-4x_1(0) + 8x_2(0) - 16x_3(0) + 1.646407\right] e^{-2t} \\
& + \left[1.5x_1(0) - 4.5x_2(0) + 13.5x_3(0) - 0.2715703\right] e^{-3t} \\
& + 1.9321005 \cos\left(0.23t - 0.3787678\right) \\
& - 0.617842 \cos\left(0.98t - 1.3837937\right) \\
& + 0.1619179 \cos\left(2.76t + 0.6620605\right) \hspace{2cm} (3.51)
\end{aligned}
$$

$$
\begin{aligned}
x_3(t) = & \left[0.5x_1(0) - 0.5x_2(0) + 0.5x_3(0) - 0.6366872\right] e^{-t} + \left[-x_1(0)\right. \\
& \left. + 2x_2(0) - 4x_3(0) + 0.4116019\right] e^{-2t} + \left[0.5x_1(0) - 1.5x_2(0)\right. \\
& \left. + 4.5x_3(0) - 0.0905234\right] e^{-3t} + 0.3217803 \cos\left(0.23t\right. \\
& \left. - 0.4170822\right) - 0.1016086 \cos\left(0.98t - 1.5466523\right) \\
& + 0.0245167 \cos\left(2.76t + 0.2309187\right) \hspace{2cm} (3.52)
\end{aligned}
$$

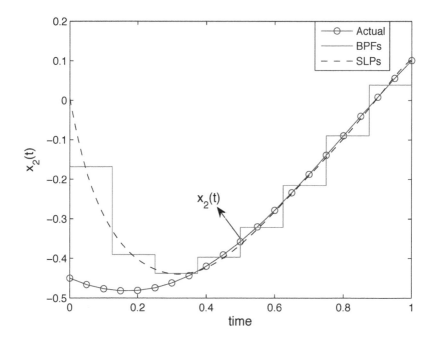

Figure 3.9: Estimate of state-2 with NSR = 0.1.

Just as in the previous example, here also, first the state estimation is done with

$$\hat{\mathbf{x}}(t_0) = \mathbf{x}(0) = \left[\; 2.7663638,\; 1.8085984,\; 0.3156087 \;\right]^T,$$

$\Delta T = 0.025$ over the interval $t \in [0,\; 3]$, and $m = 8$ on each subinterval of length unity. Figs. 3.10–3.12 show the state estimates obtained.

Next, the initial conditions $x_1(0) = 2.7663638$, $x_2(0) = 1.8085984$ and $x_3(0) = 0.3156087$ are chosen in such a manner that the transient components disappear. Assuming that $\hat{x}_1(0) = \hat{x}_2(0) = \hat{x}_3(0) = 0$, the state estimates in noise-free and noisy (NSR = 0.1) environments are obtained, and the results are given in Figs. 3.13–3.18. The results once again confirm the usefulness

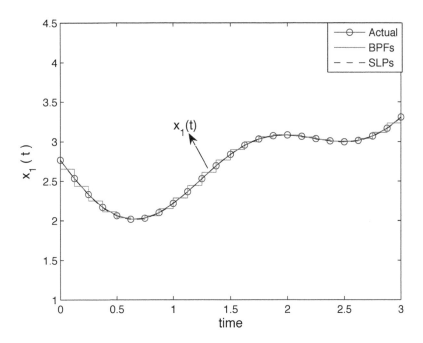

Figure 3.10: Estimate of state-1 with known initial states.

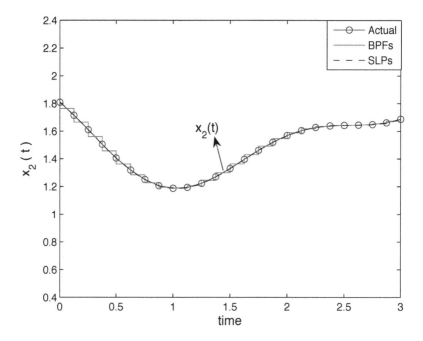

Figure 3.11: Estimate of state-2 with known initial states.

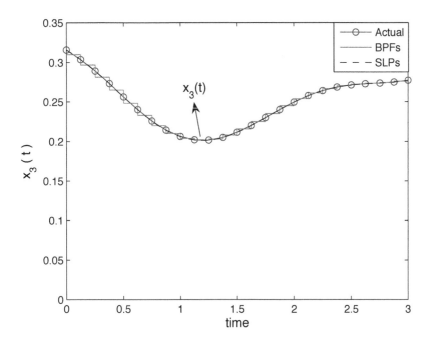

Figure 3.12: Estimate of state-3 with known initial states.

of recursive algorithms.

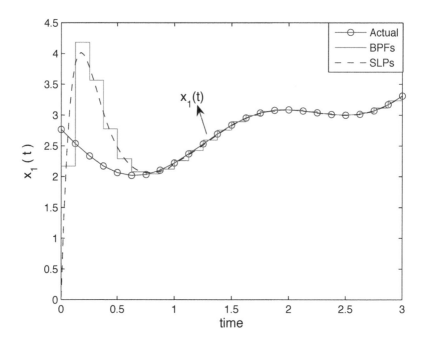

Figure 3.13: Estimate of state-1.

3.5   Conclusion

OFs (BPFs and SLPs) make state estimation possible even in the presence of noise if they are used in an estimation process using the Luenberger observer. This is due to the inherent filtering property of OFs.

The recursive algorithm in [23] is applicable to single-input single-output systems whose dynamical equations are available in observable canonical form. It is not applicable to multi-input multi-output (MIMO) systems, and systems not described in observable canonical form. This is not the case with the recursive algorithms in Sections 3.3.2 and 3.3.3 because these algorithms

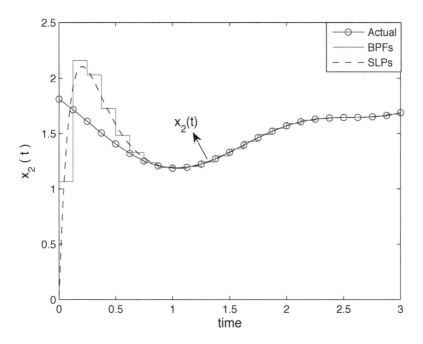

Figure 3.14: Estimate of state-2.

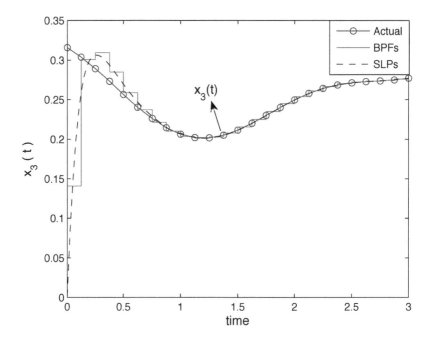

Figure 3.15: Estimate of state-3.

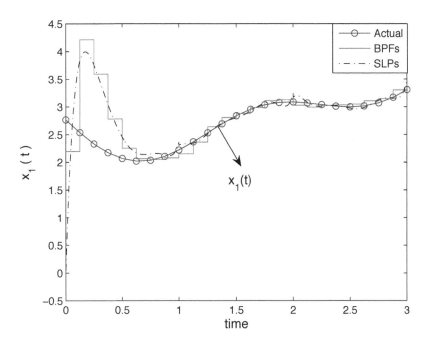

Figure 3.16: Estimate of state-1 with NSR = 0.1.

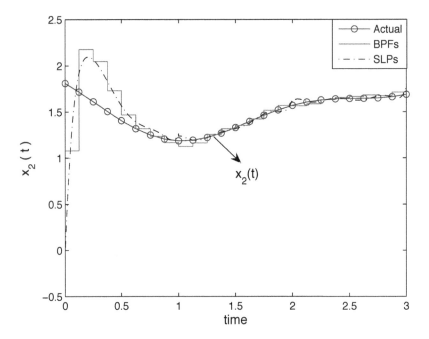

Figure 3.17: Estimate of state-2 with NSR = 0.1.

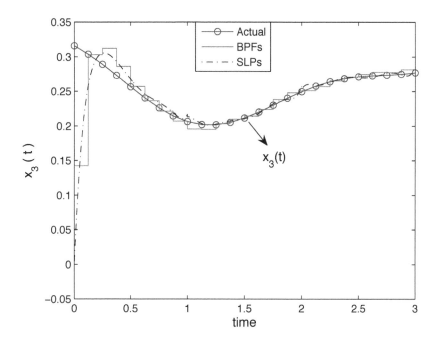

Figure 3.18: Estimate of state-3 with NSR = 0.1.

directly deal with system matrices A, B, and C. The only condition which must be satisfied to apply these algorithms is that the system should be completely observable.

Chapter 4

Linear Optimal Control Systems Incorporating Observers

Using BPFs and SLPs two recursive algorithms are presented for the analysis of linear time-invariant optimal control systems incorporating observers. An illustrative example is included to demonstrate the superiority of recursive algorithms over the non-recursive approaches.

4.1 Introduction

Consider a linear time-invariant completely observable and completely controllable system described by

$$\dot{\mathbf{x}}(t) = A\mathbf{x}(t) + B\mathbf{u}(t) \tag{4.1}$$

$$\mathbf{y}(t) = C\mathbf{x}(t) \tag{4.2}$$

where $\mathbf{u}(t)$, $\mathbf{x}(t)$ and $\mathbf{y}(t)$ are the plant input, state, and output vectors, respectively, and A, B and C are $n \times n$, $n \times r$ and $p \times n$

real, constant matrices, respectively. Assume that rank of C is p. An observer described by

$$\dot{\mathbf{z}}(t) = F\mathbf{z}(t) + G\mathbf{u}(t) + H\mathbf{y}(t) \qquad (4.3)$$
$$\hat{\mathbf{x}}(t) = L_1\mathbf{y}(t) + L_2\mathbf{z}(t) \qquad (4.4)$$

can provide the estimate $\hat{\mathbf{x}}(t)$ for the state $\mathbf{x}(t)$, where F, G, H, L_1 and L_2 are real constant $q \times q$, $q \times r$, $q \times p$, $n \times p$, and $n \times q$ matrices, respectively, and $q = n - p$, when the following conditions are satisfied [2]:

$$G = \Gamma B \qquad (4.5)$$

where Γ is the $q \times n$ matrix which is the solution of the matrix equation

$$\Gamma A - F\Gamma = HC \qquad (4.6)$$
$$\mathbf{z}(t) = \Gamma\mathbf{x}(t) + \mathbf{e}(t) \qquad (4.7)$$
$$\dot{\mathbf{e}}(t) = F\mathbf{e}(t) \qquad (4.8)$$
$$L_1 C + L_2\Gamma = I_n \qquad (4.9)$$

When an observer is incorporated to generate an estimate $\hat{\mathbf{x}}(t)$ of the plant state vector, we need to choose the matrix K in the feedback law

$$\mathbf{u}(t) = -K\hat{\mathbf{x}}(t) \qquad (4.10)$$

so that the cost function

$$J = \frac{1}{2} \int_{t_0}^{\infty} \left[\mathbf{x}^T(t) Q \mathbf{x}(t) + \mathbf{u}^T(t) R \mathbf{u}(t) \right] dt \qquad (4.11)$$

is a minimum. The $n \times n$ matrix Q and the $r \times r$ matrix R are real symmetric positive semidefinite and real symmetric positive definite, respectively.

Substituting Eqs. (4.2), (4.4), (4.7) and (4.9) into Eq. (4.10), we obtain

$$\mathbf{u}^\star(t) = -K\left[\mathbf{x}(t) + L_2\mathbf{e}(t)\right] \tag{4.12}$$

Inserting Eq. (4.12) into Eq. (4.1) yields

$$\dot{\mathbf{x}}(t) = \tilde{A}\mathbf{x}(t) + \tilde{B}\mathbf{e}(t) \tag{4.13}$$

where

$$\tilde{A} = A - BK \tag{4.14}$$

and

$$\tilde{B} = -BKL_2 \tag{4.15}$$

It follows from Eq. (4.12) that the solutions of Eqs. (4.8) and (4.13) are necessary to compute the control law $\mathbf{u}^\star(t)$.

The problem of optimal control incorporating observers has been successfully studied via different classes of OFs, namely BPFs [19], SLPs [30, 47], SJPs [33], GOPs [41], SCFs [44, 47], SCP1s [25, 47], SCP2s [47] and single-term Walsh series [52]. The approach followed in [25, 30, 33, 41, 44, 47] is non-recursive, while it is recursive in [19, 52], making the approach in [25, 30, 33, 41, 44, 47] computationally not attractive.

In this chapter, using BPFs and SLPs, two recursive algorithms [77, 79, 86] are developed for solving the problem of linear optimal control systems incorporating observers. The chapter is organized as follows: the next section deals with a unified method, also called the Kronecker product method, to solve Eqs. (4.8) and (4.13) via OFs, and discusses its demerits. Then we present two recursive algorithms using BPFs and SLPs to solve Eqs. (4.8) and (4.13). One numerical example is considered in Section 4.3 to compare the performances of recursive algorithms. Section 4.4 concludes the chapter.

4.2 Analysis of Linear Optimal Control Systems Incorporating Observers

We express the state vector $\mathbf{x}(t)$ and the error vector $\mathbf{e}(t)$ in terms of OFs as

$$\mathbf{x}(t) \approx \sum_{i=0}^{m-1} \mathbf{x}_i \phi_i(t) = X\boldsymbol{\phi}(t) \tag{4.16}$$

$$\mathbf{e}(t) \approx \sum_{i=0}^{m-1} \mathbf{e}_i \phi_i(t) = E\boldsymbol{\phi}(t) \tag{4.17}$$

where

$$X = \begin{bmatrix} \mathbf{x}_0, & \mathbf{x}_1, & \dots, & \mathbf{x}_{m-1} \end{bmatrix} \tag{4.18}$$

$$E = \begin{bmatrix} \mathbf{e}_0, & \mathbf{e}_1, & \dots, & \mathbf{e}_{m-1} \end{bmatrix} \tag{4.19}$$

which are not yet known, and $\boldsymbol{\phi}(t)$ is either a BPF vector $\mathbf{B}(t)$ or an SLP vector $\mathbf{L}(t)$. Integrating Eq. (4.8) once with respect to t and using Eqs. (2.10)/(2.50) and (4.17) yield

$$E = E_0 + FEH_f \tag{4.20}$$

where

$$
\begin{aligned}
E_0 &= \begin{bmatrix} \mathbf{e}(t_0), & \mathbf{e}(t_0), & \ldots, & \mathbf{e}(t_0) \end{bmatrix} \quad \text{if } \phi(t) = \mathbf{B}(t) \quad (4.21)\\
&= \begin{bmatrix} \mathbf{e}(t_0), & \mathbf{0}, & \mathbf{0}, & \ldots, & \mathbf{0} \end{bmatrix} \qquad \text{if } \phi(t) = \mathbf{L}(t) \quad (4.22)
\end{aligned}
$$

Similarly, from Eqs. (4.13), (2.10)/(2.50) and (4.16) we obtain

$$
X = X_0 + \left(\tilde{A}X + \tilde{B}E \right) H_f \tag{4.23}
$$

where

$$
\begin{aligned}
X_0 &= \begin{bmatrix} \mathbf{x}(t_0), & \mathbf{x}(t_0), & \ldots, & \mathbf{x}(t_0) \end{bmatrix} \quad \text{if } \phi(t) = \mathbf{B}(t) \quad (4.24)\\
&= \begin{bmatrix} \mathbf{x}(t_0), & \mathbf{0}, & \mathbf{0}, & \ldots, & \mathbf{0} \end{bmatrix} \qquad \text{if } \phi(t) = \mathbf{L}(t) \quad (4.25)
\end{aligned}
$$

Algebraic equations (4.20) and (4.23) are to be solved for the unknowns E and X. Once E and X are available, the desired control law $\mathbf{u}^\star(t)$ can be computed from Eq. (4.12) as

$$
\mathbf{u}^\star(t) = -K \left[X + L_2 E \right] \phi(t) \tag{4.26}
$$

We discuss methods of solving Eqs. (4.20) and (4.23) in the following subsections.

4.2.1 Kronecker product method

Eqs. (4.20) and (4.23) can be written as

$$
\begin{aligned}
\mathrm{Vec}(E) &= \left(I_{qm} - H_f^T \otimes F \right)^{-1} \mathrm{Vec}(E_0) \tag{4.27}\\
\mathrm{Vec}(X) &= \left(I_{nm} - H_f^T \otimes \tilde{A} \right)^{-1} \mathrm{Vec}(\tilde{V}) \tag{4.28}
\end{aligned}
$$

where

$$
\tilde{V} = X_0 + \tilde{B}EH_f = \begin{bmatrix} \tilde{\mathbf{v}}_0, & \tilde{\mathbf{v}}_1, & \ldots, & \tilde{\mathbf{v}}_{m-1} \end{bmatrix}. \tag{4.29}
$$

Vec($\cdot$) is a vector valued function of a matrix, I_{nm} is the identity matrix of order $nm \times nm$, and $\otimes$ is the Kronecker product [12] of matrices.

Though this method [77, 79, 86] is straightforward, it is not elegant computationally as it involves inversion of a matrix of size qm or nm, see Eqs. (4.27) and (4.28), which becomes larger as the value of m increases. Note that the accuracy of end result depends on m value.

4.2.2 Recursive algorithm via BPFs

In [19] $\dot{\mathbf{e}}(t)$ and $\dot{\mathbf{x}}(t)$ were also expressed in terms of BPFs to derive a recursive algorithm. Since Eqs. (4.20) and (4.23) were obtained only after integrating Eqs. (4.8) and (4.13), there is no need to express $\dot{\mathbf{e}}(t)$ and $\dot{\mathbf{x}}(t)$ in terms of BPFs to derive the recursive algorithm. Thus, the approach [77, 86] here is different from the one in [19].

Substituting the matrix H_f of BPFs into Eqs. (4.20) and (4.23) and simplifying, we obtain

$$\mathbf{e}_0 = (I_q - 0.5TF)^{-1}\,\mathbf{e}(t_0) \tag{4.30}$$

$$\mathbf{e}_i = (I_q - 0.5TF)^{-1}\,(I_q + 0.5TF)\,\mathbf{e}_{i-1} \tag{4.31}$$

for $i = 1, 2, 3, \ldots, m - 1$,

$$\mathbf{x}_0 = \left(I_n - 0.5T\tilde{A}\right)^{-1}\left[\mathbf{x}(t_0) + 0.5T\tilde{B}\mathbf{e}_0\right] \tag{4.32}$$

$$\mathbf{x}_i = \left(I_n - 0.5T\tilde{A}\right)^{-1}\left[\left(I_n + 0.5T\tilde{A}\right)\mathbf{x}_{i-1} + 0.5T\tilde{B}\left(\mathbf{e}_{i-1} + \mathbf{e}_i\right)\right] \tag{4.33}$$

for $i = 1, 2, 3, \ldots, m - 1$.

4.2.3 Recursive algorithm via SLPs

Substituting the matrix H_f of SLPs into Eq. (4.20) and rearranging the terms, we have Eq. (3.26) where

$$
W_{ij} = \begin{cases}
\frac{2I_q}{(t_f - t_0)} - F & \text{if } i = j = 0 \\[2mm]
\frac{F}{(2i+3)} & \text{if } i = 0, 1, 2, \ldots, m-2 \quad \text{and} \quad j = i+1 \\[2mm]
\frac{-F}{(2i-1)} & \text{if } i = 1, 2, 3, \ldots, m-1 \quad \text{and} \quad j = i-1 \\[2mm]
\frac{2I_q}{(t_f - t_0)} & \text{if } i = j = 1, 2, \ldots, m-1 \\[2mm]
O & \text{otherwise}
\end{cases}
$$

$$\tag{4.34}$$

$$
\mathbf{v}_i = \begin{cases}
\frac{2\mathbf{e}(t_0)}{(t_f - t_0)} & \text{if } i = 0 \\[2mm]
0 & \text{otherwise}
\end{cases}
\tag{4.35}
$$

$$
\mathbf{s}_i = \mathbf{e}_i \quad \text{for all } i \tag{4.36}
$$

Similarly, substituting the matrix H_f of SLPs into Eq. (4.23) and rearranging the terms leads to Eq. (3.26) where W_{ij} is as given in Eq. (3.27),

$$
\mathbf{v}_i = \begin{cases}
\frac{2\mathbf{x}(t_0)}{(t_f - t_0)} + \tilde{B}\left(\mathbf{e}_0 - \frac{1}{3}\mathbf{e}_1\right) & \text{if } i = 0 \\[2mm]
\tilde{B}\left(\frac{\mathbf{e}_{i-1}}{2i-1} - \frac{\mathbf{e}_{i+1}}{2i+3}\right) & \text{if } i = 1, 2, \ldots, m-2 \\[2mm]
\tilde{B}\frac{\mathbf{e}_{m-2}}{(2m-3)} & \text{if } i = m-1
\end{cases}
$$

$$\tag{4.37}$$

$$
\mathbf{s}_i = \mathbf{x}_i \quad \text{for all } i \tag{4.38}
$$

Now Eq. (3.26) with Eqs. (4.34)–(4.36) or Eqs. (3.27), (4.37) and (4.38) can be solved recursively using the recursive relations in Eqs. (3.30)–(3.34).

In Eq. (3.32) the size of the matrix to be inverted is kept to q or n instead of qm or nm as in the case of the Kronecker product method. Moreover, the matrix in Eq. (3.26) is sparse, which is fully exploited in deriving the recursive relations (3.30)–(3.34). This method [79, 86] is thus more elegant computationally.

4.3 Illustrative Example

Consider the linear system [19, 25, 30, 33, 41, 44, 47, 52]

$$\begin{bmatrix} \dot{x}_1(t) \\ \dot{x}_2(t) \end{bmatrix} = \begin{bmatrix} 0 & 1 \\ 1 & 0 \end{bmatrix} \begin{bmatrix} x_1(t) \\ x_2(t) \end{bmatrix} + \begin{bmatrix} 0 \\ -1 \end{bmatrix} u(t),$$

$$\begin{bmatrix} x_1(0) \\ x_2(0) \end{bmatrix} = \begin{bmatrix} -0.6 \\ 0.35 \end{bmatrix}$$

$$y(t) = \begin{bmatrix} 1 & 0 \end{bmatrix} \begin{bmatrix} x_1(t) & x_2(t) \end{bmatrix}^T$$

where the optimal control law is taken to be

$$u^{\star}(t) = -K\hat{\mathbf{x}}(t) = \begin{bmatrix} 1.5 & 1 \end{bmatrix} \begin{bmatrix} \hat{x}_1(t) & \hat{x}_2(t) \end{bmatrix}^T$$

due to incomplete measurement of the state, and $\hat{\mathbf{x}}(t)$ is obtained by the Luenberger observer

$$\dot{z}(t) = -1.5z(t) - u(t) - 1.25y(t), \quad z(0) = 0.5$$

$$\begin{bmatrix} \hat{x}_1(t) \\ \hat{x}_2(t) \end{bmatrix} = \begin{bmatrix} 1 \\ 1.5 \end{bmatrix} y(t) + \begin{bmatrix} 0 \\ 1 \end{bmatrix} z(t)$$

$$z(t) = \begin{bmatrix} -1.5 & 1 \end{bmatrix} \begin{bmatrix} x_1(t) & x_2(t) \end{bmatrix}^T + e(t)$$

which is designed as per the design procedure in [41] as follows: By choosing

$$L_1 = \begin{bmatrix} 1 & 1.5 \end{bmatrix}^T \quad \text{and} \quad L_2 = \begin{bmatrix} 0 & 1 \end{bmatrix}^T$$

one can obtain Γ from Eq. (4.9) as $\Gamma = \begin{bmatrix} \gamma_1 & \gamma_2 \end{bmatrix} = \begin{bmatrix} -1.5 & 1 \end{bmatrix}$.
Once Γ is obtained, G can be computed from Eq. (4.5) as $G = -1$
and F and H can be computed from Eq. (4.6) as $F = -1.5$ and
$H = -1.25$. Therefore, one can have

$$A = \begin{bmatrix} 0 & 1 \\ 1 & 0 \end{bmatrix}, \quad B = \begin{bmatrix} 0 \\ -1 \end{bmatrix}, \quad C = \begin{bmatrix} 1 & 0 \end{bmatrix}, \quad K = \begin{bmatrix} -1.5 & -1 \end{bmatrix},$$

$$F = -1.5, G = -1, \quad H = -1.25, \quad L_1 = \begin{bmatrix} 1 \\ 1.5 \end{bmatrix},$$

$$L_2 = \begin{bmatrix} 0 \\ 1 \end{bmatrix}, \quad \Gamma = \begin{bmatrix} -1.5 & 1 \end{bmatrix},$$

$$\tilde{A} = A - BK = \begin{bmatrix} 0 & 1 \\ -0.5 & -1 \end{bmatrix}, \quad \text{and} \quad \tilde{B} = -BKL_2 = \begin{bmatrix} 0 \\ -1 \end{bmatrix}$$

On each unit interval $m = 4$ is considered and $e(t)$, $\mathbf{x}(t)$, and
$u^\star(t)$ over $t \in [0, 5]$ are computed using both the recursive algorithms. Fig. 4.1 shows exact

$$e(t) = -0.75\, e^{-1.5t}$$

and $e(t)$ obtained via SLPs and BPFs. Fig. 4.2 shows $x_1(t)$ and
$x_2(t)$ obtained via SLPs and BPFs while Fig. 4.3 shows exact

$$u^\star(t) = -0.75\, e^{-1.5t} + e^{-0.5t}\, [1.9 \sin(0.5t) - 0.55 \cos(0.5t)]$$

and $u^\star(t)$ obtained via SLPs and BPFs. It can be observed from
Figs. 4.1–4.3 that BPF approach always produces a piecewise constant solution and the SLP approach produces a continuous solution.

The results obtained by the recursive algorithms in Sections
4.2.2 and 4.2.3 are in close agreement with the exact results. Moreover, the same results obtained via the non-recursive SLP approach
in [30] and the recursive BPF approach in [19] are also included

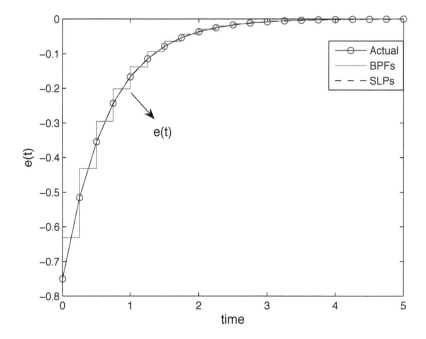

Figure 4.1: Error.

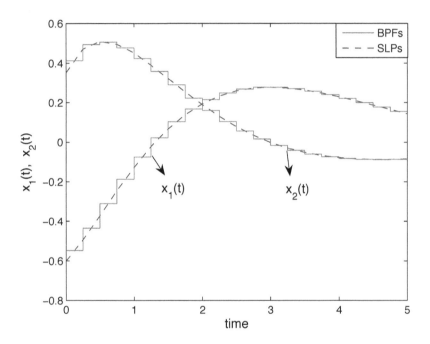

Figure 4.2: State variables.

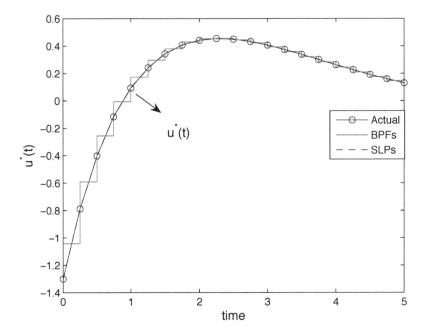

Figure 4.3: Optimal control law.

for comparison's sake. The SLP approach in Section 4.2.3 is faster and less complex than the SLP approach in [30], as it is now purely recursive in nature.

4.4 Conclusion

Based on using BPFs and SLPs, two recursive algorithms are presented for analysis of linear optimal control systems incorporating observers. Computational superiority of these algorithms over the algorithms reported in the literature has been discussed. For a fixed value of m, the number of SLPs or BPFs, the BPF algorithm is faster than the SLP algorithm but its end result is piecewise constant; not smooth as in the case of SLP algorithm. So, one has a choice to choose a method (BPF or SLP) based on the requirement, i.e. speed of computation or accuracy of end result.

Chapter 5

Optimal Control of Systems Described by Integro-Differential Equations

In this chapter, a unified approach via BPFs or SLPs is presented to solve the optimal control problem of linear time-invariant systems described by integro-differential equations. By using the elegant operational properties of OFs (BPFs or SLPs), computationally elegant algorithms are developed for calculating optimal control law and state trajectory of dynamical systems. A numerical example is included to demonstrate the validity of the approach.

5.1 Introduction

Synthesis of optimal control law for deterministic systems described by the integro-differential equations has been investigated [11] via the dynamic programming approach. Subsequently, this problem has been studied via OF approach. In [37] BPFs were used, while in [34, 43] SLPs were applied to study the problem. Moreover, SCP1s were employed in [42]. All these methods dealt with time-varying systems in the development of computational algorithms. In this chapter, we consider time-invariant systems, introduce a unified approach [80] via BPFs or SLPs to solve the optimal control problem of such systems described by integro-differential equations, and demonstrate the validity of the approach by considering the same numerical example in [37, 34, 42, 43].

The chapter is organized as follows: The next section deals with the optimal control problem. A numerical example is included in Section 5.3. The last section concludes the chapter.

5.2 Optimal Control of LTI Systems Described by Integro-Differential Equations

Consider the linear time-invariant system described by

$$\dot{\mathbf{x}}(t) = A\mathbf{x}(t) + B\mathbf{u}(t) + \int_{t_0}^{t_f} G(t, \tau)\mathbf{x}(\tau) \, d\tau \qquad (5.1)$$

with $\mathbf{x}(t_0)$ specified where $\mathbf{x}(t) \in R^n$ is the state vector, $\mathbf{u}(t) \in R^r$ is the control vector, A and $G(t, \tau)$ are the matrices of order $n \times n$, B is the matrix of order $n \times r$, and t_0, t_f are initial and final times, respectively. $G(t, \tau) = 0$ for $t < \tau$ and the elements of G are assumed to be bounded and continuous in the interval $[t_0, t_f]$.

The problem is to find $\mathbf{u}(t)$ which minimizes the cost function

$$J = \frac{1}{2} \int_{t_0}^{t_f} \left[\mathbf{x}^T(t) \, Q \, \mathbf{x}(t) + \mathbf{u}^T(t) R \, \mathbf{u}(t) \right] dt \qquad (5.2)$$

where Q is an $n \times n$ real symmetric positive semidefinite matrix and R is an $r \times r$ real symmetric positive definite matrix.

Integrate Eq. (5.1) with respect to t to obtain

$$\mathbf{x}(t) - \mathbf{x}(t_0) = \int_{t_0}^{t} \left\{ A\mathbf{x}(\sigma) + B\mathbf{u}(\sigma) + \int_{t_0}^{t_f} G(\sigma,\, \tau)\mathbf{x}(\tau)\, d\tau \right\} d\sigma$$

(5.3)

Express $\mathbf{x}(t)$, $\mathbf{x}(t_0)$, $\mathbf{u}(t)$ and $G(t,\, \tau)$ in terms of OFs $\{\phi_i(t)\}$, which can be BPFs or SLPs, as follows:

$$\mathbf{x}(t) \approx \sum_{i=0}^{m-1} \mathbf{x}_i \phi_i(t) = X\boldsymbol{\phi}(t) \tag{5.4}$$

$$\mathbf{x}(t_0) = X_0 \boldsymbol{\phi}(t) \tag{5.5}$$

$$\mathbf{u}(t) \approx \sum_{i=0}^{m-1} \mathbf{u}_i \phi_i(t) = U\boldsymbol{\phi}(t) \tag{5.6}$$

$$G(t,\, \tau) \approx \sum_{i=0}^{m-1} \sum_{j=0}^{m-1} G_{ij}\, \phi_i(t)\, \phi_j(\tau) \tag{5.7}$$

where

$$X = \begin{bmatrix} \mathbf{x}_0, & \mathbf{x}_1, & \ldots, & \mathbf{x}_{m-1} \end{bmatrix} \tag{5.8}$$

$$X_0 = \begin{bmatrix} \mathbf{x}(t_0), & \mathbf{x}(t_0), & \ldots, & \mathbf{x}(t_0) \end{bmatrix} \quad \text{if} \quad \boldsymbol{\phi}(t) = \mathbf{B}(t) \tag{5.9}$$

$$= \begin{bmatrix} \mathbf{x}(t_0), & \mathbf{0}, & \mathbf{0}, & \ldots\ldots, & \mathbf{0} \end{bmatrix} \quad \text{if} \quad \boldsymbol{\phi}(t) = \mathbf{L}(t) \tag{5.10}$$

$$U = \begin{bmatrix} \mathbf{u}_0, & \mathbf{u}_1, & \ldots, & \mathbf{u}_{m-1} \end{bmatrix} \tag{5.11}$$

$$G_{ij} = \frac{1}{T^2} \int_{t_0+iT}^{t_0+(i+1)T} \int_{t_0+jT}^{t_0+(j+1)T} G(t,\, \tau)\, dt\, d\tau \quad \text{if BPFs used} \tag{5.12}$$

$$= \frac{(2i+1)(2j+1)}{(t_f - t_0)^2} \int_{t_0}^{t_f} \int_{t_0}^{t_f} G(t,\, \tau) L_i(t) L_j(\tau) dt\, d\tau \quad \text{if SLPs used}$$

(5.13)

Now we can write

$$\int_{t_0}^{t_f} G(t,\tau)\mathbf{x}(\tau)\,d\tau = \int_{t_0}^{t_f} \sum_{i=0}^{m-1}\sum_{j=0}^{m-1} G_{ij}\,\phi_i(t)\,\phi_j(\tau) \sum_{k=0}^{m-1}\mathbf{x}_k\phi_k(\tau)d\tau$$

$$= \sum_{i=0}^{m-1}\sum_{j=0}^{m-1}\sum_{k=0}^{m-1} G_{ij}\mathbf{x}_k\,\phi_i(t)\int_{t_0}^{t_f}\phi_j(\tau)\phi_k(\tau)d\tau$$

$$= W\boldsymbol{\phi}(t) \tag{5.14}$$

where

$$W = \begin{bmatrix} \mathbf{w}_0 & \mathbf{w}_1 & \cdots & \mathbf{w}_{m-1} \end{bmatrix} \tag{5.15}$$

$$\mathbf{w}_i = T\sum_{j=0}^{m-1} G_{ij}\mathbf{x}_j \quad \text{if BPFs used} \tag{5.16}$$

$$= (t_f - t_0)\sum_{j=0}^{m-1} \frac{G_{ij}\mathbf{x}_j}{2j+1} \quad \text{if SLPs used} \tag{5.17}$$

Substitute Eqs. (5.4)–(5.6) and (5.14) into Eq. (5.3) and make use of the integration operational property in Eq. (2.10) or (2.50) to have

$$X - X_0 = (AX + BU + W)H$$

$$\Rightarrow \quad \hat{\mathbf{x}} - \hat{\mathbf{x}}_0 = \left(H^T \otimes A\right)\hat{\mathbf{x}} + \left(H^T \otimes B\right)\hat{\mathbf{u}} + \left(H^T \otimes I_n\right)\hat{\mathbf{w}} \tag{5.18}$$

where $\otimes$ is the Kronecker product [12],

$$\hat{\mathbf{x}} = \begin{bmatrix} \mathbf{x}_0 \\ \mathbf{x}_1 \\ \vdots \\ \mathbf{x}_{m-1} \end{bmatrix}, \hat{\mathbf{u}} = \begin{bmatrix} \mathbf{u}_0 \\ \mathbf{u}_1 \\ \vdots \\ \mathbf{u}_{m-1} \end{bmatrix}, \hat{\mathbf{x}}_0 = \begin{bmatrix} \mathbf{x}(t_0) \\ \mathbf{x}(t_0) \\ \vdots \\ \mathbf{x}(t_0) \end{bmatrix} \text{ or } \begin{bmatrix} \mathbf{x}(t_0) \\ \mathbf{0} \\ \vdots \\ \mathbf{0} \end{bmatrix},$$

$$\hat{\mathbf{w}} = \begin{bmatrix} \mathbf{w}_0 \\ \mathbf{w}_1 \\ \vdots \\ \mathbf{w}_{m-1} \end{bmatrix} \tag{5.19}$$

$$\hat{\mathbf{w}} = \hat{G}\hat{\mathbf{x}} \tag{5.20}$$

and

$$\hat{G} = T \begin{bmatrix} G_{00} & G_{01} & \cdots & G_{0,m-1} \\ G_{10} & G_{11} & \cdots & G_{1,m-1} \\ \vdots & \vdots & & \vdots \\ G_{m-1,0} & G_{m-1,1} & \cdots & G_{m-1,m-1} \end{bmatrix} \quad \text{if BPFs used} \tag{5.21}$$

$$= (t_f - t_0) \begin{bmatrix} G_{00} & \frac{G_{01}}{3} & \cdots & \frac{G_{0,m-1}}{(2m-1)} \\ G_{10} & \frac{G_{11}}{3} & \cdots & \frac{G_{1,m-1}}{(2m-1)} \\ \vdots & \vdots & & \vdots \\ G_{m-1,0} & \frac{G_{m-1,1}}{3} & \cdots & \frac{G_{m-1,m-1}}{(2m-1)} \end{bmatrix} \quad \text{if SLPs used} \tag{5.22}$$

Consequently Eq. (5.18) can be written as

$$\hat{\mathbf{x}} = M\hat{\mathbf{u}} + \hat{\mathbf{v}} \tag{5.23}$$

where

$$M = M_1 \left(H^T \otimes B \right) \tag{5.24}$$

$$M_1 = \left[I_{mn} - \left(H^T \otimes A \right) - \left(H^T \otimes I_n \right) \hat{G} \right]^{-1} \tag{5.25}$$

and

$$\hat{\mathbf{v}} = M_1 \hat{\mathbf{x}}_0. \tag{5.26}$$

Similarly, the cost function in Eq. (5.2) becomes

$$J = \frac{1}{2} \left[\hat{\mathbf{x}}^T \hat{Q} \hat{\mathbf{x}} + \hat{\mathbf{u}}^T \hat{R} \hat{\mathbf{u}} \right] \tag{5.27}$$

where

$$\hat{Q} = P \otimes Q, \quad \hat{R} = P \otimes R \tag{5.28}$$

and

$$P = T \times \text{diag}\begin{bmatrix} 1, & 1, & \ldots, & 1 \end{bmatrix} \quad \text{if BPFs used} \qquad (5.29)$$

$$= (t_f - t_0) \times \text{diag}\begin{bmatrix} 1, & \frac{1}{3}, & \ldots, & \frac{1}{(2m-1)} \end{bmatrix} \quad \text{if SLPs are used}$$

$$(5.30)$$

Substituting Eq. (5.23) into Eq. (5.27) and setting the optimization condition

$$\partial J / \partial \hat{u} = \mathbf{0}^T \qquad (5.31)$$

yield the optimal control law

$$\hat{\mathbf{u}} = -\left(M^T \hat{Q} M + \hat{R}\right)^{-1} M^T \hat{Q} \hat{\mathbf{v}}. \qquad (5.32)$$

5.3   Illustrative Example

Consider the motion of the controlled system described [11, 37, 34, 42, 43] by

$$\dot{x}(t) = x(t) + u(t) + \int_0^1 g(t, \tau) x(\tau) d\tau$$

with

$$g(t, \tau) = \begin{cases} 2 - 4(t - \tau) & \text{for } 0 \le (t - \tau) \le 0.5 \\ 0 & \text{for } 0 > (t - \tau) > 0.5 \end{cases}$$

and $x(0) = 1$. The cost function is given by

$$J = \frac{1}{2} \int_0^1 \left[x^2(t) + 2u^2(t) \right] dt$$

The optimal control law $u(t)$ and the state variable $x(t)$ are computed with $m = 4$ and 32 for BPF approach and $m = 4$ for SLP approach. The results obtained are shown in Figs. 5.1 and 5.2. It is clear from the figures that the results of the BPF approach and SLP approach follow each other. Table 5.1 shows the J value for different m values in the BPF approach and for $m = 4$ in the SLP approach.

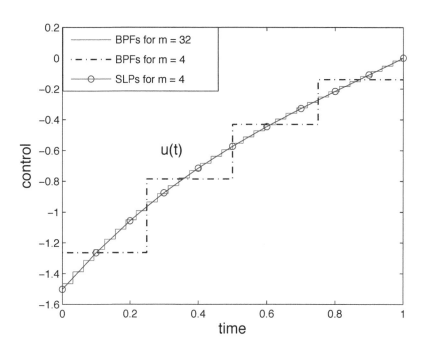

Figure 5.1: BPF and SLP solutions of control variable.

Table 5.1: Cost function

Method	m	J
BPF	4	1.5451
	8	1.5136
	16	1.5058
	24	1.5044
	32	1.5039
SLP	4	1.5032

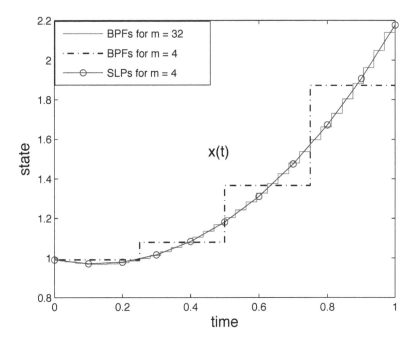

Figure 5.2: BPF and SLP solutions of state variable.

5.4   Conclusion

In this chapter a unified approach is presented to solve linear quadratic optimal control problem of time-invariant deterministic systems described by integro-differential equations. In the example considered, the results obtained with $m = 4$ SLPs are quite satisfactory. Since BPFs are piecewise constant and the actual solution is smooth, one has to choose a large value for m in BPF approach in order to improve upon the accuracy.

Chapter 6

Linear-Quadratic-Gaussian Control

In this chapter, a unified approach and two recursive algorithms via BPFs and SLPs are presented to solve the LQG control problem. By using the elegant operational properties of OFs (BPFs or SLPs), computationally elegant algorithms are developed. A numerical example is included to demonstrate the validity of the unified approach and recursive algorithms.

6.1 Introduction

The LQG control problem [4] concerns linear systems disturbed by additive white Gaussian noise, incomplete state information and quadratic costs. The LQG controller is simply the combination of a LQE, i.e., a Kalman filter with a LQR. The separation principle guarantees that these can be designed and computed independently.

In [35] the solution of the LQG control design problem has been obtained by employing GOPs. By using the GOPs the nonlinear Riccati differential equations have been reduced to nonlinear algebraic equations. The set of nonlinear algebraic equations has been solved to get the solutions. The above approach is neither simple nor attractive computationally, as nonlinear equations are involved.

As it appears from the literature the LQG control problem has not been attempted via other classes of OFs, i.e. piecewise constant orthogonal functions. Hence in this chapter, we consider linear time-invariant systems and introduce two new methods to solve the LQG control problem of such systems using BPFs or SLPs. We develop a unified approach and recursive algorithms to solve the LQG control problem.

The chapter is organized as follows: The next section deals with the LQG control design problem. The method of obtaining a solution of the LQG control design problem is presented in Sections 6.3 and 6.4 which contain a unified approach and two recursive algorithms, respectively. A numerical example is included and demonstrates the effectiveness of unified approach and recursive algorithms. The last section concludes the chapter.

6.2   LQG Control Problem

Consider the linear dynamic system

$$\dot{\mathbf{x}}(t) = A\mathbf{x}(t) + B\mathbf{u}(t) + \mathbf{v}(t) \tag{6.1}$$

$$\mathbf{z}(t) = C\mathbf{x}(t) + \mathbf{w}(t) \tag{6.2}$$

where $\mathbf{x}$ represents an n - vector of state variables, $\mathbf{u}$ a p - vector of control variables, and $\mathbf{z}$ a q - vector of measured outputs, and $\mathbf{v}$ and $\mathbf{w}$ the additive zero-mean white Gaussian system noise and additive zero-mean white Gaussian measurement noise, respectively, i.e.

$$\mathrm{E}\left\{\mathbf{v}(t)\mathbf{v}^T(\tau)\right\} \;=\; Q_2\,\delta(t-\tau) \tag{6.3}$$

$$\mathrm{E}\left\{\mathbf{w}(t)\mathbf{w}^T(\tau)\right\} \;=\; R_2\,\delta(t-\tau) \tag{6.4}$$

where Q_2 is nonnegative definite symmetric and R_2 is positive definite symmetric. Also $\mathbf{v}(t)$ is uncorrelated with $\mathbf{w}(t)$, i.e.

$$\mathrm{E}\left\{\mathbf{v}(t)\mathbf{w}^T(\tau)\right\} \;=\; 0 \tag{6.5}$$

Assume that the initial state vector $\mathbf{x}(t_0)$ is Gaussian with mean $\bar{\mathbf{x}}(t_0)$ and covariance matrix

$$P \;=\; P_2(t_0) \;=\; \mathrm{E}\left\{[\mathbf{x}(t_0)-\bar{\mathbf{x}}(t_0)]\,[\mathbf{x}(t_0)-\bar{\mathbf{x}}(t_0)]^T\right\}$$

which is symmetric nonnegative definite, and

$$\mathrm{E}\left\{\mathbf{v}(t)\,\mathbf{x}_0^T\right\} \;=\; \mathrm{E}\left\{\mathbf{w}(t)\,\mathbf{x}_0^T\right\} \;=\; 0 \quad \text{for } t \geq t_0 \tag{6.6}$$

Given this system, the objective is to find the control input $\mathbf{u}(t)$, which at every time t may depend only on the past measurements $\mathbf{z}(t_1)$, $t_0 \leq t_1 < t$ such that the cost function

$$J \;=\; \mathrm{E}\left\{\frac{1}{2}\mathbf{x}^T(t_f)S\mathbf{x}(t_f) + \frac{1}{2}\int_{t_0}^{t_f}\left[\mathbf{x}^T(t)Q_1\mathbf{x}(t) + \mathbf{u}^T(t)R_1\mathbf{u}(t)\right]dt\right\} \tag{6.7}$$

is minimized, where the matrix R_1 is positive definite symmetric, and S and Q_1 are nonnegative definite symmetric.

The LQG controller that solves the LQG control problem is specified by the equations

$$\dot{\hat{\mathbf{x}}}(t) \;=\; A\hat{\mathbf{x}}(t) + B\mathbf{u}(t) + K_2(t)\left[\mathbf{z}(t) - C\hat{\mathbf{x}}(t)\right] \tag{6.8}$$

$$\hat{\mathbf{x}}(t_0) \;=\; \mathrm{E}\left[\mathbf{x}(t_0)\right] \;=\; \bar{\mathbf{x}}_0$$

$$\mathbf{u}(t) \;=\; -K_1(t)\hat{\mathbf{x}}(t) \tag{6.9}$$

The matrix $K_2(t)$ is called the Kalman gain of the associated Kalman filter represented by Eq. (6.8). At each time t this filter generates estimates $\hat{\mathbf{x}}(t)$ of the state $\mathbf{x}(t)$ using the past measurements and inputs. The Kalman gain is determined through the associated matrix Riccati differential equation

$$\dot{P}_2(t) = AP_2(t) + P_2(t)A^T - P_2(t)C^T R_2^{-1} C P_2(t) + Q_2 \quad (6.10)$$
$$P_2(t_0) = P$$

Given the solution $P_2(t)$, $t_0 \le t \le t_f$ the Kalman gain equals

$$K_2(t) = P_2(t)C^T R_2^{-1} \qquad (6.11)$$

The matrix $K_1(t)$ is called the feedback gain matrix, which is determined through the associated matrix Riccati differential equation

$$-\dot{P}_1(t) = A^T P_1(t) + P_1(t)A - P_1(t)BR_1^{-1}B^T P_1(t) + Q_1 \quad (6.12)$$
$$P_1(t_f) = S$$

Given the solution $P_1(t)$, $t_0 \le t \le t_f$ the feedback gain equals

$$K_1(t) = R_1^{-1}B^T P_1(t) \qquad (6.13)$$

Observe the similarity of the two matrix Riccati differential Equations (6.10) and (6.12); the first one running forward in time, and the second one running backward in time. The first one solves the LQE problem and the second one solves the LQR problem. So the LQG control problem separates into LQE and LQR problems that can be solved independently.

The block diagram [10] of the LQG control problem is presented in Figure 6.1.

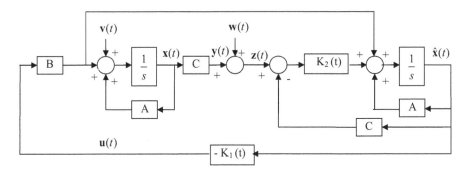

Figure 6.1: Optimum linear combined estimation and control.

6.3 Unified Approach

Integrating the Riccati equation (6.12) backward in time from t_f to t, we obtain

$$- [P_1(t) - S] = \int_{t_f}^{t} \left[A^T P_1(\tau) + P_1(\tau)A - P_1(\tau)F P_1(\tau) + Q_1 \right] d\tau$$

(6.14)

where $F = BR_1^{-1}B^T$. Expressing $P_1(t)$, $P_1(t)FP_1(t)$, Q_1 and S in terms of OFs $\{\phi_i(t)\}$, which may be BPFs $\{B_i(t)\}$ or SLPs $\{L_i(t)\}$, we have

$$P_1(t) \simeq \sum_{i=0}^{m-1} P_{1i}\phi_i(t) = \tilde{P}_1 \left(\phi(t) \otimes I_n \right),$$

(6.15)

where

$$\tilde{P}_1 = \left[\begin{array}{cccc} P_{10}, & P_{11}, & \ldots, & P_{1,m-1} \end{array} \right].$$

(6.16)

Then

$$A^T P_1(t) \simeq \bar{P}_1 \left(\phi(t) \otimes I_n \right)$$

(6.17)

$$P_1(t)A \simeq \hat{P}_1 \left(\phi(t) \otimes I_n \right)$$

(6.18)

where

$$\bar{P}_1 = \left[\begin{array}{cccc} A^T P_{10}, & A^T P_{11}, & \ldots, & A^T P_{1,m-1} \end{array} \right] \tag{6.19}$$

$$\hat{P}_1 = \left[\begin{array}{cccc} P_{10} A, & P_{11} A, & \ldots, & P_{1,m-1} A \end{array} \right] \tag{6.20}$$

$$P_1(t) F P_1(t) \simeq \sum_{i=0}^{m-1} P_{1i} F P_{1i} \, \phi_i(t) \quad \text{if } \boldsymbol{\phi}(t) \text{ is } \mathbf{B}(t) \tag{6.21}$$

$$\simeq \sum_{i=0}^{m-1} \sum_{j=0}^{m-1} \sum_{k=0}^{m-1} P_{1i} F P_{1j} \, \psi_{ijk} \, \phi_k(t) \quad \text{if } \boldsymbol{\phi}(t) \text{ is } \mathbf{L}(t) \tag{6.22}$$

$$\simeq \tilde{F} \left(\boldsymbol{\phi}(t) \otimes I_n \right) \tag{6.23}$$

where

$$\tilde{F} = \left[\begin{array}{cccc} P_{10} F P_{10}, & P_{11} F P_{11}, & \ldots, & P_{1,m-1} F P_{1,m-1} \end{array} \right] \text{ if } \boldsymbol{\phi}(t) \text{ is } \mathbf{B}(t) \tag{6.24}$$

$$= \left[\sum_{i=0}^{m-1} \sum_{j=0}^{m-1} \psi_{ij0} P_{1i} F P_{1j}, \quad \sum_{i=0}^{m-1} \sum_{j=0}^{m-1} \psi_{ij1} P_{1i} F P_{1j}, \quad \ldots \ldots \right.$$

$$\left. \ldots \ldots, \quad \sum_{i=0}^{m-1} \sum_{j=0}^{m-1} \psi_{ij,m-1} P_{1i} F P_{1j} \right] \text{ if } \boldsymbol{\phi}(t) \text{ is } \mathbf{L}(t) \tag{6.25}$$

$$Q_1 = \tilde{Q}_1 \left(\boldsymbol{\phi}(t) \otimes I_n \right) \tag{6.26}$$

where

$$\tilde{Q}_1 = \left[\begin{array}{cccc} Q_1, & Q_1, & \ldots, & Q_1 \end{array} \right] \text{ if } \boldsymbol{\phi}(t) \text{ is } \mathbf{B}(t) \tag{6.27}$$

$$= \left[\begin{array}{cccc} Q_1, & 0, & \ldots \ldots, & 0 \end{array} \right] \text{ if } \boldsymbol{\phi}(t) \text{ is } \mathbf{L}(t) \tag{6.28}$$

and

$$S = \tilde{S} \left(\boldsymbol{\phi}(t) \otimes I_n \right) \tag{6.29}$$

where

$$\tilde{S} = [\, S, \ S, \ \ldots, \ S \,] \quad \text{if } \phi(t) \text{ is } \mathbf{B}(t) \tag{6.30}$$

$$= [\, S, \ 0, \ \ldots, \ 0 \,] \quad \text{if } \phi(t) \text{ is } \mathbf{L}(t) \tag{6.31}$$

and $\otimes$ is the Kronecker product [12] of matrices.

Substituting Eqs. (6.15), (6.17), (6.18), (6.23), (6.26) and (6.29) into Eq. (6.14) and making use of the backward integration operational property in Eq. (2.12) or (2.52), we have

$$-\tilde{P}_1 + \tilde{S} = \left[\bar{P}_1 + \hat{P}_1 - \tilde{F} + \tilde{Q}_1 \right] (H_b \otimes I_n)$$

$$\Rightarrow \quad \tilde{P}_1 + \left[\bar{P}_1 + \hat{P}_1 - \tilde{F} \right] (H_b \otimes I_n) = \tilde{S} - \tilde{Q}_1 (H_b \otimes I_n) \tag{6.32}$$

which is to be solved for the spectrum of $P_1(t)$. Similarly, the spectrum of $P_2(t)$ can also be found from the Riccati equation (6.10), and is given by

$$\tilde{P}_2 - \left[\bar{P}_2 + \hat{P}_2 - \tilde{G} \right] (H_f \otimes I_n) = \tilde{P} + \tilde{Q}_2 (H_f \otimes I_n) \tag{6.33}$$

where

$$\bar{P}_2 = [\, AP_{20}, \ AP_{21}, \ \ldots, \ AP_{2,m-1} \,] \tag{6.34}$$

$$\hat{P}_2 = [\, P_{20}A^T, \ P_{21}A^T, \ \ldots, \ P_{2,m-1}A^T \,] \tag{6.35}$$

and $G = C^T R_2^{-1} C$

Notice that both the Riccati equations are thus reduced to the nonlinear algebraic equations which can be solved using the Newton-Raphson method.

6.3.1 Illustrative example

Consider the linear system [10, 35]

$$\dot{x}(t) = -0.5x(t) + u(t) + v(t)$$
$$\bar{x}(0) = 10$$

with the measurement

$$z(t) = x(t) + w(t)$$

and the cost function

$$J = E\left\{0.5x^2(t_f)S + 0.5\int_0^{t_f}\left[2x^2(t) + u^2(t)\right]dt\right\}$$

where

$$E\{v(t)v(\tau)\} = 2\,\delta(t-\tau)$$
$$E\{w(t)w(\tau)\} = 0.25\,\delta(t-\tau)$$
$$E\{[x(0) - \bar{x}(0)]^2\} = 0$$

If $S = 0$ and $t_f = 1$, the exact solutions of $P_1(t)$ and $P_2(t)$ are given by

$$P_1(t) = -0.5 + 1.5\tanh(-1.5t + 1.8465736) \quad \text{and}$$
$$P_2(t) = -0.125 + 0.125\sqrt{33}\tanh\left\{0.5\sqrt{33}\,t + \tan^{-1}\left(1/\sqrt{33}\right)\right\}$$

Now this problem is to be solved by using BPFs and SLPs. As the given system is a scalar system it is possible to obtain a recursive algorithm if BPFs are used. This point is discussed here. Substituting the operational matrix of backward integration H_b in Eq. (2.13) into Eq. (6.32) and simplifying, we obtain the following recursive algorithm :

$$P_{1,m-1} = -\frac{1}{F}\left(\frac{1}{T} - A\right) + \sqrt{\left[\frac{1}{F}\left(\frac{1}{T} - A\right)\right]^2 + \frac{Q_1}{F}}$$

$$P_{1,j} = -\frac{1}{F}\left(\frac{1}{T} - A\right)$$

$$+\sqrt{\left[\frac{1}{F}\left(\frac{1}{T} - A\right)\right]^2 + \frac{2Q_1}{F} + \frac{2}{F}\left(\frac{1}{T} + A\right)P_{1,j+1} - P_{1,j+1}^2}$$

for $j = m - 2, m - 3, \ldots, 1, 0$.

Similarly, from Eqs. (2.11) and (6.33) we have

$$P_{2,0} = -\frac{1}{G}\left(\frac{1}{T} - A\right) + \sqrt{\left[\frac{1}{G}\left(\frac{1}{T} - A\right)\right]^2 + \frac{Q_2}{G}}$$

$$P_{2,j} = -\frac{1}{G}\left(\frac{1}{T} - A\right)$$

$$+\sqrt{\left[\frac{1}{G}\left(\frac{1}{T} - A\right)\right]^2 + \frac{2Q_2}{G} + \frac{2}{G}\left(\frac{1}{T} + A\right)P_{2,j-1} - P_{2,j-1}^2}$$

for $j = 1, 2, \ldots, m - 1$.

Such a recursive algorithm is not possible with SLPs. So with $m = 4$ and 24 the above recursive algorithm via BPFs and with $m = 4$ the nonrecursive approach via SLPs are applied, and $P_1(t)$ and $P_2(t)$ are computed as shown in Figs. 6.2 and 6.3. The actual solutions are also shown in the same figures for comparison. The results are quite satisfactory even with four SLPs.

6.4 Recursive Algorithms

The nonlinear matrix Riccati differential equation (6.10) can be written as two linear equations [7] as

$$\dot{Y}_2(t) = -A^T Y_2(t) + F_2 Z_2(t) \qquad (6.36)$$

$$\dot{Z}_2(t) = Q_2 Y_2(t) + A Z_2(t) \qquad (6.37)$$

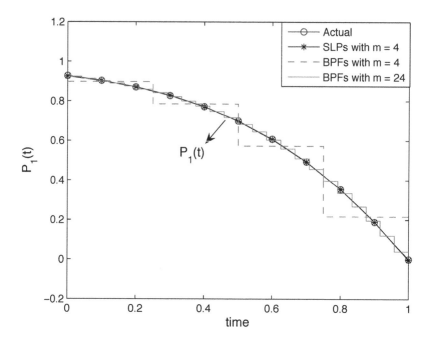

Figure 6.2: Actual, SLP and BPF solutions of $P_1(t)$.

where $Y_2(t)$ is an $n \times n$ nonsingular matrix, $Z_2(t)$ an $n \times n$ matrix and

$$F_2 = C^T R_2^{-1} C \tag{6.38}$$

$$P_2(t) = Z_2(t) Y_2^{-1}(t), \quad t_0 \le t \le t_f. \tag{6.39}$$

Equations (6.36) and (6.37) can be written as

$$\dot{W}_2(t) = G_2 W_2(t) \tag{6.40}$$

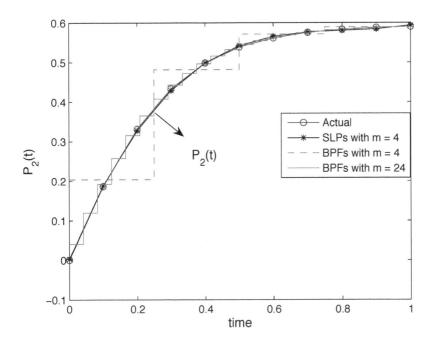

Figure 6.3: Actual, SLP and BPF solutions of $P_2(t)$.

where

$$W_2(t) = \begin{bmatrix} Y_2(t) \\ \cdots\cdots \\ Z_2(t) \end{bmatrix} \tag{6.41}$$

$$G_2 = \begin{bmatrix} -A^T & \vdots & F_2 \\ \cdots\cdots & \vdots & \cdots\cdots \\ Q_2 & \vdots & A \end{bmatrix} \tag{6.42}$$

Integrating Eq. (6.40) from t_0 to t yields

$$W_2(t) - W_2(t_0) = G_2 \int_{t_0}^{t} W_2(\tau)d\tau \tag{6.43}$$

Expressing $W_2(t)$ and $W_2(t_0)$ in terms of OFs $\{\phi_i(t)\}$, which can be BPFs or SLPs, as follows:

$$W_2(t) \approx \sum_{i=0}^{m-1} W_{2i}\phi_i(t) = \tilde{W}_2\left(\phi(t) \otimes I_n\right) \qquad (6.44)$$

$$W_2(t_0) = \tilde{W}_{20}\left(\phi(t) \otimes I_n\right) \qquad (6.45)$$

where

$$\tilde{W}_2 = \begin{bmatrix} W_{20}, & W_{21}, & \dots, & W_{2,m-1} \end{bmatrix} \qquad (6.46)$$

$$\tilde{W}_{20} = \begin{bmatrix} W_2(t_0), & W_2(t_0), & \dots, & W_2(t_0) \end{bmatrix} \quad \text{if } \phi(t) = \mathbf{B}(t) \ (6.47)$$

$$= \begin{bmatrix} W_2(t_0), & O, & O, & \dots, & O \end{bmatrix} \quad \text{if } \phi(t) = \mathbf{L}(t) \ (6.48)$$

Substituting Eqs. (6.44) and (6.45) into Eq. (6.43) and making use of forward integration operational property in Eq. (2.10) or (2.50), we have

$$\tilde{W}_2 - \tilde{W}_{20} = G_2\tilde{W}_2\left(H_f \otimes I_n\right) \qquad (6.49)$$

which is to be solved for $\tilde{W}_2$. We discuss two recursive methods for solving Eq. (6.49) in the following subsections.

In a similar manner the non-linear Riccati Eq. (6.12) can be written as

$$\dot{Y}_1(t) = AY_1(t) - F_1Z_1(t) \qquad (6.50)$$

$$\dot{Z}_1(t) = -Q_1Y_1(t) - A^T Z_1(t) \qquad (6.51)$$

where $Y_1(t)$ is an $n \times n$ nonsingular matrix, $Z_1(t)$ an $n \times n$ matrix and

$$F_1 = BR_1^{-1}B^T \qquad (6.52)$$

$$P_1(t) = Z_1(t)Y_1^{-1}(t), \quad t_0 \leq t \leq t_f. \qquad (6.53)$$

From Eqs. (6.50) and (6.51) we can write

$$\dot{W}_1(t) = G_1 W_1(t) \tag{6.54}$$

where

$$W_1(t) = \begin{bmatrix} Y_1(t) \\ \cdots\cdots \\ Z_1(t) \end{bmatrix} \tag{6.55}$$

$$G_1 = \begin{bmatrix} A & \vdots & -F_1 \\ \cdots\cdots & \vdots & \cdots\cdots \\ -Q_1 & \vdots & -A^T \end{bmatrix} \tag{6.56}$$

6.4.1 Recursive algorithm via BPFs

Substituting the matrix H_f in Eq. (2.11) and matrix $\tilde{W}_{20}$ in Eq. (6.47) into Eq. (6.49) and simplifying, we get

$$W_{20} = (I_{2n} - 0.5TG_2)^{-1} W_2(t_0) \tag{6.57}$$

$$W_{2i} = (I_{2n} - 0.5TG_2)^{-1} (I_{2n} + 0.5TG_2) W_{2,i-1} \tag{6.58}$$

for $i = 1, 2, 3, \ldots, m - 1$.

Once $W_2(t)$ (i.e. $Y_2(t)$ and $Z_2(t)$) is known, the spectrum of $P_2(t)$ can be calculated easily from Eq. (6.39).

In a similar manner, the spectrum of $W_1(t)$ can be found recursively by first integrating Eq. (6.54) from t_f to t and then using the backward operational matrix of integration in Eq. (2.13). This leads to

$$W_{1,m-1} = (I_{2n} + 0.5TG_1)^{-1} W_1(t_f) \tag{6.59}$$

$$W_{1i} = (I_{2n} + 0.5TG_1)^{-1} (I_{2n} - 0.5TG_1) W_{1,i+1} \tag{6.60}$$

for $i = m - 2, m - 3, \ldots, 1, 0$.

Now the spectrum of $P_1(t)$ can be calculated from Eq. (6.53).

6.4.2 Recursive algorithm via SLPs

Substituting the matrix H_f in Eq. (2.51) and matrix $\tilde{W}_{20}$ in Eq. (6.48) into Eq. (6.49) and rearranging the terms, we have

$$\tilde{G}\tilde{W} = \tilde{V} \tag{6.61}$$

where

$$\tilde{G} = \begin{bmatrix} G_{00} & G_{01} & \bigcirc & \bigcirc & \cdots & & \bigcirc \\ G_{10} & G_{11} & G_{12} & \bigcirc & \cdots & & \bigcirc \\ \bigcirc & G_{21} & G_{22} & G_{23} & \cdots & & \bigcirc \\ \vdots & \vdots & \vdots & \vdots & & & \vdots \\ \bigcirc & \bigcirc & \bigcirc & \bigcirc & \cdots & G_{m-2,\,m-1} \\ \bigcirc & \bigcirc & \bigcirc & \bigcirc & \cdots & G_{m-1,\,m-1} \end{bmatrix} ;$$

$$\tilde{W} = \begin{bmatrix} W_0 \\ W_1 \\ W_2 \\ \vdots \\ W_{m-2} \\ W_{m-1} \end{bmatrix} ; \tilde{V} = \begin{bmatrix} V_0 \\ \bigcirc \\ \bigcirc \\ \vdots \\ \bigcirc \\ \bigcirc \end{bmatrix} \tag{6.62}$$

with

$$G_{ij} = \begin{cases} I_{2n} - \Gamma G_2 & \text{if } i = j = 0 \\[2mm] \frac{\Gamma G_2}{(2i+3)} & \text{if } i = 0, 1, 2, \ldots, m-2 \text{ and } j = i+1 \\[2mm] \frac{-\Gamma G_2}{(2i-1)} & \text{if } i = 1, 2, 3, \ldots, m-1 \text{ and } j = i-1 \\[2mm] I_{2n} & \text{if } i = j = 1, 2, \ldots, m-1 \\[2mm] \bigcirc_{2n} & \text{otherwise} \end{cases} \tag{6.63}$$

$$V_0 = W_2(t_0) = \begin{bmatrix} I_n \\ P \end{bmatrix} \tag{6.64}$$

$$W_i = W_{2i} \quad \text{for all } i \tag{6.65}$$

and

$$\Gamma \; = \; \frac{(t_f - t_0)}{2} \tag{6.66}$$

Similarly by integrating Eq. (6.54) from t_f to t and using backward operational matrix of integration in Eq. (2.53) and rearranging the terms lead to Eq. (6.62) where

$$G_{ij} \; = \; \begin{cases} I_{2n} + \Gamma G_1 & \text{if } i = j = 0 \\[2mm] \frac{\Gamma G_1}{(2i+3)} & \text{if } i = 0, 1, 2, \ldots, m - 2 \text{ and } j = i + 1 \\[2mm] \frac{-\Gamma G_1}{(2i-1)} & \text{if } i = 1, 2, 3, \ldots, m - 1 \text{ and } j = i - 1 \\[2mm] I_{2n} & \text{if } i = j = 1, 2, \ldots, m - 1 \\[2mm] O_{2n} & \text{otherwise} \end{cases} \tag{6.67}$$

$$V_0 \; = \; W_1(t_f) \; = \; \begin{bmatrix} I_n \\ S \end{bmatrix} \tag{6.68}$$

$$W_i \; = \; W_{1i} \quad \text{for all } i \tag{6.69}$$

Now Eq. (6.62) with Eqs. (6.63)–(6.65) or Eqs. (6.67)–(6.69) can be solved recursively using the following recursive relations:

$$R_{i,i-1} \; = \; -M_{ii}G_{i,i-1} \quad \text{for } i = m - 1, m - 2, \ldots, 2, 1. \tag{6.70}$$

$$M_{ii} \; = \; \begin{cases} G_{ii}^{-1} & \text{if } i = m - 1 \\[2mm] (G_{ii} + G_{i,i+1}R_{i+1,i})^{-1} & \text{if } i = m - 2, m - 3, \ldots, 1, 0. \end{cases} \tag{6.71}$$

$$W_0 \; = \; M_{00}V_0 \tag{6.72}$$

$$W_i \; = \; R_{i,i-1}W_{i-1} \quad \text{for } i = 1, 2, \ldots, m - 1. \tag{6.73}$$

Once $W_2(t)$ and $W_1(t)$ are known, $P_2(t)$ and $P_1(t)$ can be found from Eqs. (6.39) and (6.53), respectively.

6.4.3 Illustrative example

Consider the same example as given in Section 6.3.1. Now the problem is solved by using the above recursive algorithms via BPFs and SLPs. $P_1(t)$ and $P_2(t)$ are computed with $m = 6$ and shown in Figs. 6.4 and 6.5. The actual solutions are also shown in the same figures. The results are precisely matching, even with six SLPs. For obtaining improved results with BPFs, next $m = 60$ is considered and shown in Figs. 6.4 and 6.5.

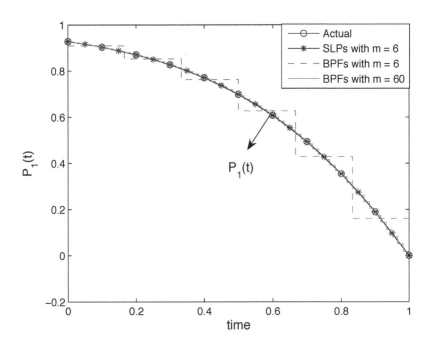

Figure 6.4: Actual, SLP and BPF solutions of $P_1(t)$.

The error (difference between the actual solution and the computed solution with $m = 6$ SLPs) in the recursive approach and the nonrecursive approach [35] is shown in Figs 6.6 and 6.7 for comparison. From the figures it is clear that the results obtained with recursive algorithms are more accurate.

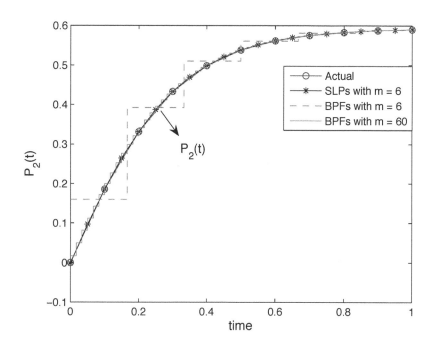

Figure 6.5: Actual, SLP and BPF solutions of $P_2(t)$.

6.5   Conclusion

A unified method and two recursive algorithms are introduced to determine the filter gain and the regulator gain in the LQG control problem. An illustrative example is included to demonstrate the usefulness of the unified approach and the recursive algorithms via SLPs and BPFs. One has to consider a large number of BPFs to improve upon the accuracy. This is because we are using piecewise constant functions (BPFs) to represent the smooth functions $P_1(t)$ and $P_2(t)$ in the present context. It is clear from Figs. 6.6 and 6.7 that the results obtained by the recursive algorithm are more accurate than what was obtained in [35].

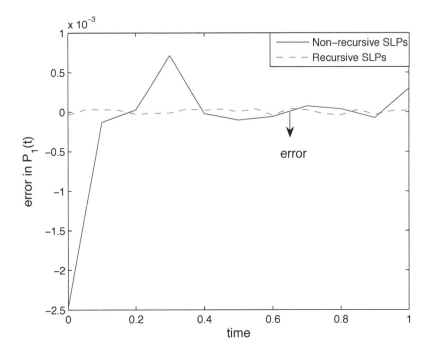

Figure 6.6: Error in $P_1(t)$: recursive SLPs and non-recursive SLPs [35].

Every approach (SLP or BPF) has its own advantage and disadvantage. The SLP method does not require a large number of polynomials in a series expansion to represent smooth functions, but computationally it is not as attractive as the BPF method because SLPs are to be computed and used for signal representation while it is not so in BPF method, as BPFs are all unity.

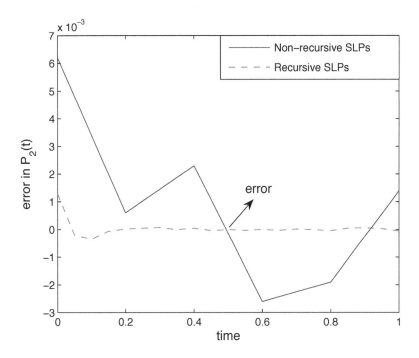

Figure 6.7: Error in $P_2(t)$: recursive SLPs and non-recursive SLPs [35].

Chapter 7

Optimal Control of Singular Systems

In this chapter, two recursive algorithms are presented for computing the optimal control law of linear time-invariant singular systems with quadratic performance index by using the elegant properties of BPFs and SLPs. Also, a unified approach is given to solve the optimal control problem of singular systems via BPFs or SLPs. Two numerical examples are included to demonstrate the validity of the recursive algorithms and unified approach.

7.1 Introduction

Singular systems have been of considerable importance as they are often encountered in many areas. Singular systems arise naturally in describing large-scale systems [53]; examples occur in power and interconnected systems. In general, an interconnection of state variable subsystems is conveniently described as a singular system. The singular system is called a generalized state-space system, implicit system, semi-state system, or descriptor system. The optimal regulator problem of such systems can be stated as follows:

Consider the linear time-invariant system

$$E\dot{\mathbf{x}}(t) = A\mathbf{x}(t) + B\mathbf{u}(t) \tag{7.1}$$

$$\mathbf{x}(t_0) = \tilde{\mathbf{x}}_0, \quad \mathbf{x}(t_f) = \mathbf{x}_f \tag{7.2}$$

where $\mathbf{x}(t) \in R^n$ is the state vector, $\mathbf{u}(t) \in R^r$ is the input vector, E is an $n \times n$ singular matrix, and A and B are $n \times n$ and $n \times r$ constant matrices, respectively.

The problem is to find $\mathbf{u}(t)$ that will drive the system in Eq. (7.1) from an initial state $\mathbf{x}(t_0)$ to a fixed final state $\mathbf{x}(t_f)$ in a fixed time interval $(t_f - t_0)$ while minimizing the cost function

$$J = \frac{1}{2} \int_{t_0}^{t_f} \left[\mathbf{x}^T(t) Q \mathbf{x}(t) + \mathbf{u}^T(t) R \mathbf{u}(t) \right] dt \tag{7.3}$$

where Q and R are real symmetric positive semidefinite matrix and real symmetric positive definite matrix, respectively.

The above problem has been discussed in [15] and [18]. In [38] the necessary conditions for the existence of optimal controls have been derived. They have shown that the optimal control design problem reduces to a two-point boundary-value (TPBV) problem for the determination of the optimal state trajectory. A single-term Walsh series method [56] has been applied to study the optimal control problem of singular systems. In [62] SLPs were used to solve the same problem. However, this approach is nonrecursive in nature. The Haar wavelet approach [68] has been presented to study the optimal control problem of linear singularly perturbed systems. In the recent times, SCFs [71], SCP1s [74] and Legendre wavelets [82] have been applied for solving the optimal control problem of singular systems. These approaches are again nonrecursive.

In this chapter, using two classes of OFs (BPFs and SLPs) two recursive algorithms [85, 92] are introduced to solve the optimal control problem of singular systems. These algorithms are computationally elegant in view of their recursive nature. A unified approach [85, 92] is also proposed for solving the optimal control problem via BPFs or SLPs.

The chapter is organized as follows: a solution to the problem considered is discussed in Sections 7.2 and 7.3 which contain two recursive algorithms and a unified approach, respectively. Two illustrative examples are included in Section 7.4. The last section concludes the chapter.

7.2 Recursive Algorithms

Assume that [38]

$$E = \begin{bmatrix} E_1 \\ E_2 \end{bmatrix} = \begin{bmatrix} I_{n-r} & \bigcirc \\ E_{21} & E_{22} \end{bmatrix} \quad \text{and} \quad B = \begin{bmatrix} \bigcirc \\ I_r \end{bmatrix} \quad (7.4)$$

where I_{n-r} and I_r are the identity matrices of order $n - r$ and r respectively, $\bigcirc$ is the null matrix of appropriate order, and E_{21} and E_{22} are $r \times (n - r)$ and $r \times r$ matrices respectively. Partition the matrices A, Q, the state $\mathbf{x}(t)$ and the co-state $\mathbf{p}(t)$ as follows:

$$A = \begin{bmatrix} A_{11} & A_{12} \\ A_{21} & A_{22} \end{bmatrix}, \quad Q = \begin{bmatrix} Q_{11} & Q_{12} \\ Q_{12}^T & Q_{22} \end{bmatrix},$$

$$\mathbf{x}(t) = \begin{bmatrix} \mathbf{x}_1(t) \\ \mathbf{x}_2(t) \end{bmatrix} \quad \text{and} \quad \mathbf{p}(t) = \begin{bmatrix} \mathbf{p}_1(t) \\ \mathbf{p}_2(t) \end{bmatrix} \quad (7.5)$$

where A_{11} & Q_{11}, A_{12} & Q_{12}, A_{21} & Q_{12}^T, A_{22} & Q_{22} are respectively $(n - r) \times (n - r)$, $(n - r) \times r$, $r \times (n - r)$, $n \times n$ matrices, $\mathbf{x}_1(t)$ and $\mathbf{p}_1(t)$ are $(n - r)-$ dimensional vectors, and $\mathbf{x}_2(t)$ and $\mathbf{p}_2(t)$ are $r-$ dimensional vectors.

Then we have the following equations [38]:

$$
\begin{bmatrix}
I & O & O & O \\
E_{21} & E_{22} & O & O \\
O & O & -E_{21}^T R & I \\
O & O & -E_{22}^T R & O
\end{bmatrix}
\begin{bmatrix}
\dot{\mathbf{x}}_1(t) \\
\dot{\mathbf{x}}_2(t) \\
\dot{\mathbf{u}}(t) \\
\dot{\mathbf{p}}_1(t)
\end{bmatrix}
=
$$

$$
\begin{bmatrix}
A_{11} & A_{12} & O & O \\
A_{21} & A_{22} & I & O \\
-Q_{11} & -Q_{12} & A_{21}^T R & -A_{11}^T \\
-Q_{12}^T & -Q_{22} & A_{22}^T R & -A_{12}^T
\end{bmatrix}
\begin{bmatrix}
\mathbf{x}_1(t) \\
\mathbf{x}_2(t) \\
\mathbf{u}(t) \\
\mathbf{p}_1(t)
\end{bmatrix}
\tag{7.6}
$$

which is to be solved with the help of two-point boundary values.

The above problem can be solved as an initial value problem if $E_{21} = E_{22} = O$ because, in this case we have

$$
\mathbf{u}(t) = -A_{21}\mathbf{x}_1(t) - A_{22}\mathbf{x}_2(t)
$$

$$
A_{12}^T\mathbf{p}_1(t) = -Q_{12}^T\mathbf{x}_1(t) - Q_{22}\mathbf{x}_2(t) + A_{22}^T R\mathbf{u}(t)
$$

and thereby $\mathbf{u}(t_0)$ and $A_{12}^T\mathbf{p}_1(t_0)$ from the knowledge of $\mathbf{x}(t_0)$.

How to solve Eq. (7.6) using OFs is discussed [85, 92] now. Eq. (7.6) can be written in the form

$$
G\dot{\mathbf{s}}(t) = F\mathbf{s}(t) \tag{7.7}
$$

where

$$
G =
\begin{bmatrix}
I & O & O & O \\
O & O & O & O \\
O & O & O & I \\
O & O & O & O
\end{bmatrix}, \quad
F =
\begin{bmatrix}
A_{11} & A_{12} & O & O \\
A_{21} & A_{22} & I & O \\
-Q_{11} & -Q_{12} & A_{21}^T R & -A_{11}^T \\
-Q_{12}^T & -Q_{22} & A_{22}^T R & -A_{12}^T
\end{bmatrix},
$$

$$
\mathbf{s}(t) =
\begin{bmatrix}
\mathbf{x}_1(t) \\
\mathbf{x}_2(t) \\
\mathbf{u}(t) \\
\mathbf{p}_1(t)
\end{bmatrix}
\tag{7.8}
$$

Integrating Eq. (7.7) with respect to t, we have

$$
G\left[\mathbf{s}(t) - \mathbf{s}(t_0)\right] = \int_{t_0}^{t} F\mathbf{s}(\tau)d\tau \tag{7.9}
$$

Now expressing $\mathbf{s}(t)$ and $\mathbf{s}(t_0)$ in terms of OFs $\{\phi_i(t)\}$ (BPFs or SLPs), we have

$$\mathbf{s}(t) \approx \sum_{i=0}^{m-1} \mathbf{s}_i \phi_i(t) = S\boldsymbol{\phi}(t) \tag{7.10}$$

$$\mathbf{s}(t_0) = S_0 \boldsymbol{\phi}(t) \tag{7.11}$$

where

$$\boldsymbol{\phi}(t) = \left[\phi_0(t), \ \phi_1(t), \ \ldots, \ \phi_{m-1}(t) \right]^T \tag{7.12}$$

$$S = \left[\mathbf{s}_0, \ \mathbf{s}_1, \ \ldots, \ \mathbf{s}_{m-1} \right] \tag{7.13}$$

$$S_0 = \left[\mathbf{s}(t_0), \ \mathbf{s}(t_0), \ \ldots, \ \mathbf{s}(t_0) \right] \quad \text{if} \quad \boldsymbol{\phi}(t) = \mathbf{B}(t) \tag{7.14}$$

$$= \left[\mathbf{s}(t_0), \ \mathbf{0}, \ \mathbf{0}, \ \ldots\ldots, \ \mathbf{0} \right] \quad \text{if} \quad \boldsymbol{\phi}(t) = \mathbf{L}(t) \tag{7.15}$$

Substituting Eqs. (7.10) and (7.11) into Eq. (7.9) and using the forward integration operational property in Eq. (2.10) or (2.50) yields

$$G(S - S_0) = FSH_f$$

$$\Rightarrow \quad GS - FSH_f = GS_0 \tag{7.16}$$

which is to be solved for S. We discuss two recursive methods in the following subsections for solving Eq. (7.16).

7.2.1 Recursive algorithm via BPFs

Substituting the matrix H_f in Eq. (2.11) and matrix S_0 in Eq. (7.14) into Eq. (7.16) and simplifying, we get [85, 92]

$$\mathbf{s}_0 = (G - 0.5TF)^{-1} G\mathbf{s}(t_0) \tag{7.17}$$

$$\mathbf{s}_i = (G - 0.5TF)^{-1} (G + 0.5TF)\,\mathbf{s}_{i-1} \tag{7.18}$$

for $i = 1, 2, 3, \ldots, m - 1$.

7.2.2 Recursive algorithm via SLPs

Substituting the matrix H_f in Eq. (2.51) and matrix S_0 in Eq. (7.15) into Eq. (7.16), and rearranging the terms, we have Eq. (3.26) where

$$
W_{ij} = \begin{cases}
\frac{2}{(t_f - t_0)} G - F & \text{if } i = j = 0 \\[2mm]
\frac{F}{(2i+3)} & \text{if } i = 0, 1, 2, \ldots, m-2 \text{ and } j = i+1 \\[2mm]
\frac{-F}{(2i-1)} & \text{if } i = 1, 2, 3, \ldots, m-1 \text{ and } j = i-1 \\[2mm]
\frac{2G}{(t_f - t_0)} & \text{if } i = j = 1, 2, 3, \ldots, m-1 \\[2mm]
O & \text{otherwise}
\end{cases}
\tag{7.19}
$$

and

$$
\mathbf{v}_i = \begin{cases}
\frac{2}{(t_f - t_0)} G\mathbf{s}(t_0) & \text{if } i = 0 \\[2mm]
\mathbf{0} & \text{otherwise}
\end{cases}
\tag{7.20}
$$

Now Eq. (3.26) with Eqs. (7.19) and (7.20) can be solved recursively [85, 92] using the following recursive relations:

$$
M_{i,m-1} = \begin{cases}
W_{ii} R_{i+1,m-1} + W_{i,i+1} R_{i+2,m-1} & \text{if } i = m-2, \ldots, 1, 0 \\
W_{ii} & \text{if } i = m-1
\end{cases}
\tag{7.21}
$$

$$
R_{i,m-1} = \begin{cases}
-W_{i,i-1}^{-1} M_{i,m-1} & \text{if } i = m-1, m-2, \ldots, 2, 1 \\
I_{2n} & \text{if } i = m
\end{cases}
\tag{7.22}
$$

$$
\mathbf{s}_{m-1} = M_{0,m-1}^{-1} \mathbf{v_0}
\tag{7.23}
$$

$$
\mathbf{s}_i = R_{i+1,m-1} \mathbf{s}_{m-1}
\tag{7.24}
$$

for $i = 0, 1, 2, \ldots, m-2$.

In Eqs. (7.22) and (7.23) the size of the matrix to be inverted is kept to $2n$ only.

7.3   Unified Approach

The recursive algorithms in the previous section can not be applied if $E_{21} = E_{22} \neq \bigcirc$. Under this situation, the following unified method [85, 92] will be helpful. The merit of this method is that it does not impose any restrictions on the matrices E, A and B.

Integrating Eq. (7.1) with respect to t, we obtain

$$E\left[\mathbf{x}(t) - \mathbf{x}(t_0)\right] = \int_{t_0}^{t} \left[A\mathbf{x}(\tau) + B\mathbf{u}(\tau)\right] d\tau \qquad (7.25)$$

Express $\mathbf{x}(t)$, $\mathbf{x}(t_0)$ and $\mathbf{u}(t)$ in terms of orthogonal functions $\{\phi_i(t)\}$, which can be BPFs or SLPs, as follows:

$$\mathbf{x}(t) \approx \sum_{i=0}^{m-1} \mathbf{x}_i \phi_i(t) = X\boldsymbol{\phi}(t) \qquad (7.26)$$

$$\mathbf{x}(t_0) = X_0 \boldsymbol{\phi}(t) \qquad (7.27)$$

$$\mathbf{u}(t) \approx \sum_{i=0}^{m-1} \mathbf{u}_i \phi_i(t) = U\boldsymbol{\phi}(t) \qquad (7.28)$$

where

$$X = \left[\begin{array}{cccc} \mathbf{x}_0, & \mathbf{x}_1, & \ldots, & \mathbf{x}_{m-1} \end{array}\right] \qquad (7.29)$$

$$X_0 = \left[\begin{array}{cccc} \mathbf{x}(t_0), & \mathbf{x}(t_0), & \ldots, & \mathbf{x}(t_0) \end{array}\right] \quad \text{if} \quad \boldsymbol{\phi}(t) = \mathbf{B}(t) \quad (7.30)$$

$$= \left[\begin{array}{cccc} \mathbf{x}(t_0), & \mathbf{0}, & \mathbf{0}, & \ldots, \mathbf{0} \end{array}\right] \quad \text{if} \quad \boldsymbol{\phi}(t) = \mathbf{L}(t) \quad (7.31)$$

Substituting Eqs. (7.26)–(7.28) into Eq. (7.25) and making use of forward integration operational property in Eq. (2.10) or (2.50), we have

$$E(X - X_0) = (AX + BU)H_f$$

$$\Rightarrow \left(I_m \otimes E - H_f^T \otimes A\right) \hat{\mathbf{x}} = (I_m \otimes E)\hat{\mathbf{x}}_0 + \left(H^T \otimes B\right) \hat{\mathbf{u}}$$

$$\Rightarrow \hat{\mathbf{x}} = M\hat{\mathbf{u}} + \hat{\mathbf{w}} \qquad (7.32)$$

where

$$M = N \left(H_f^T \otimes B \right) \tag{7.33}$$

$$N = \left(I_m \otimes E - H_f^T \otimes A \right)^{-1} \tag{7.34}$$

$$\hat{\mathbf{w}} = N \left(I_m \otimes E \right) \hat{\mathbf{x}}_0 \tag{7.35}$$

$\otimes$ is the Kronecker product of matrices [12], and

$$\hat{\mathbf{x}} = \begin{bmatrix} \mathbf{x}_0 \\ \mathbf{x}_1 \\ \vdots \\ \mathbf{x}_{m-1} \end{bmatrix}, \quad \hat{\mathbf{x}}_0 = \begin{bmatrix} \mathbf{x}(t_0) \\ \mathbf{x}(t_0) \\ \vdots \\ \mathbf{x}(t_0) \end{bmatrix} \quad \text{or} \quad \begin{bmatrix} \mathbf{x}(t_0) \\ 0 \\ \vdots \\ 0 \end{bmatrix}, \quad \hat{\mathbf{u}} = \begin{bmatrix} \mathbf{u}_0 \\ \mathbf{u}_1 \\ \vdots \\ \mathbf{u}_{m-1} \end{bmatrix} \tag{7.36}$$

Now considering the cost function in Eq. (7.3), and expressing it in terms of orthogonal functions, we have

$$\begin{aligned} J &= \frac{1}{2} \int_{t_0}^{t_f} \boldsymbol{\phi}^T(t) \left[X^T Q X + U^T R U \right] \boldsymbol{\phi}(t) dt \\ &= \frac{1}{2} \left(\hat{\mathbf{x}}^T \hat{Q} \hat{\mathbf{x}} + \hat{\mathbf{u}}^T \hat{R} \hat{\mathbf{u}} \right) \end{aligned} \tag{7.37}$$

where

$$\hat{Q} = P \otimes Q, \quad \hat{R} = P \otimes R \tag{7.38}$$

and

$$\begin{aligned} P &= T \times \text{diag} \begin{bmatrix} 1, & 1, & \ldots, & 1 \end{bmatrix} \quad \text{for} \quad \text{BPFs} \tag{7.39} \\ &= (t_f - t_0) \times \text{diag} \begin{bmatrix} 1, & \frac{1}{3}, & \ldots, & \frac{1}{(2m-1)} \end{bmatrix} \quad \text{for SLPs} \tag{7.40} \end{aligned}$$

Substituting Eq. (7.32) into Eq. (7.37) and setting the optimization condition

$$\frac{\partial J}{\partial \hat{\mathbf{u}}} = \mathbf{0}^T \tag{7.41}$$

yield the following optimal control law

$$\hat{\mathbf{u}} = - \left(M^T \hat{Q} M + \hat{R} \right)^{-1} M^T \hat{Q} \hat{\mathbf{w}} \tag{7.42}$$

7.4 Illustrative Examples

Two examples are considered here to illustrate the recursive algorithms and the unified approach via BPFs and SLPs.

Example 7.1

For the singular system [18, 38, 56, 62, 71, 82]

$$\begin{bmatrix} 0 & 1 \\ 0 & 0 \end{bmatrix} \begin{bmatrix} \dot{x}_1(t) \\ \dot{x}_2(t) \end{bmatrix} = \begin{bmatrix} 1 & 0 \\ 0 & 1 \end{bmatrix} \begin{bmatrix} x_1(t) \\ x_2(t) \end{bmatrix} + \begin{bmatrix} 0 \\ 1 \end{bmatrix} u(t) \quad (7.43)$$

find optimal control $u(t)$ that will drive the system from the initial state $\mathbf{x}(0) = \begin{bmatrix} 1, & \frac{-1}{\sqrt{2}} \end{bmatrix}^T$ to the desired final state $\mathbf{x}(t \to \infty) = \begin{bmatrix} 0, & 0 \end{bmatrix}^T$ while minimizing the cost function

$$J = \frac{1}{2} \int_0^\infty \left[\mathbf{x}^T(t)\mathbf{x}(t) + u^2(t) \right] dt. \quad (7.44)$$

In order to use the recursive algorithms in Section 7.2, we introduce the transformation

$$\begin{bmatrix} z_1(t) \\ z_2(t) \end{bmatrix} = \begin{bmatrix} 0 & 1 \\ 1 & 0 \end{bmatrix} \begin{bmatrix} x_1(t) \\ x_2(t) \end{bmatrix} \quad (7.45)$$

to obtain

$$\begin{bmatrix} 1 & 0 \\ 0 & 0 \end{bmatrix} \begin{bmatrix} \dot{z}_1(t) \\ \dot{z}_2(t) \end{bmatrix} = \begin{bmatrix} 0 & 1 \\ 1 & 0 \end{bmatrix} \begin{bmatrix} z_1(t) \\ z_2(t) \end{bmatrix} + \begin{bmatrix} 0 \\ 1 \end{bmatrix} u(t) \quad (7.46)$$

$$J = \frac{1}{2} \int_0^\infty \left[\mathbf{z}^T(t)\mathbf{z}(t) + u^2(t) \right] dt \quad (7.47)$$

The exact solution of this problem is given by

$$x_1(t) = e^{-\sqrt{2}t}, \quad x_2(t) = \frac{-1}{\sqrt{2}} e^{-\sqrt{2}t} \quad (7.48)$$

and

$$u(t) = \frac{1}{\sqrt{2}} e^{-\sqrt{2}t} \quad (7.49)$$

118

Considering the recursive algorithm in Section 7.2, we have from Eqs. (7.46) and (7.47)

$$E_{21} = E_{22} = 0, \quad A_{11} = A_{22} = 0, \quad A_{12} = A_{21} = 1,$$
$$Q_{11} = Q_{22} = 1, \quad Q_{12} = Q_{12}^T = 0.$$

Moreover, $u(t) = -z_1(t)$ and $p_1(t) = -z_2(t)$ from which we have $u(0) = -z_1(0) = \frac{1}{\sqrt{2}}$ and $p_1(0) = -z_2(0) = -1$. The recursive algorithms in the subsections 7.2.1 and 7.2.2 are applied with $m = 12$ BPFs and $m = 6$ SLPs over $t \in [0, 4]$ and the results are shown in Figures 7.1 and 7.2 which also show the actual solutions for comparison. The J value obtained via BPFs/SLPs is less than the actual value. It looks surprising. The reason for this is, the J value is optimized over $t \in [0, \infty)$ in the exact method while it is done over $t \in [0, 4]$ in BPF/SLP approach.

Table 7.1: Cost function

Method	m	J
Actual		0.35354907566247
BPF	12	0.35354644274457
SLP	6	0.35354472716328

Next, the unified approach in Section 7.3 is applied with m values remaining the same as above and obtaining exactly the same results as shown in Figures 7.1 and 7.2. The J value in each case is given in Table 7.1.

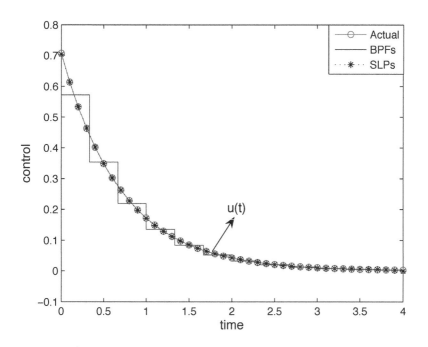

Figure 7.1: Actual, SLP and BPF solutions of $u(t)$.

Example 7.2

Consider the system described by

$$
\begin{bmatrix} 2.3425 & 0 & 1.4644 \\ 0 & 0 & 0 \\ 3.8654 & 0 & 1.4453 \end{bmatrix} \begin{bmatrix} \dot{x}_1(t) \\ \dot{x}_2(t) \\ \dot{x}_3(t) \end{bmatrix} = \begin{bmatrix} 3.3657 & 0.3563 & -1.8445 \\ 0.5823 & 1.4763 & 4.7343 \\ -1.7623 & 0.2355 & 2.7433 \end{bmatrix} \begin{bmatrix} x_1(t) \\ x_2(t) \\ x_3(t) \end{bmatrix}
$$

$$
+ \begin{bmatrix} 1.4874 & 0.4207 \\ 1.0442 & 2.3464 \\ 1.5642 & 0.4768 \end{bmatrix} \begin{bmatrix} u_1(t) \\ u_2(t) \end{bmatrix} \quad (7.50)
$$

$$
\begin{bmatrix} x_1(0) \\ x_2(0) \\ x_3(0) \end{bmatrix} = \begin{bmatrix} 1 \\ 1 \\ 1 \end{bmatrix} \quad (7.51)
$$

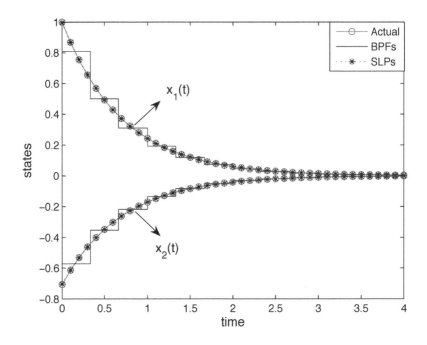

Figure 7.2: Actual, SLP and BPF solutions of $\mathbf{x}(t)$.

to find the optimal control law $\mathbf{u}(t)$ which minimizes the cost function

$$
J = \frac{1}{2} \int_0^1 \left\{ \begin{bmatrix} x_1(t) & x_2(t) & x_3(t) \end{bmatrix} \begin{bmatrix} 0.5473 & 1.4369 & 0.5634 \\ 1.4369 & 1.4465 & 0.2467 \\ 0.5634 & 0.2467 & 0.4567 \end{bmatrix} \begin{bmatrix} x_1(t) \\ x_2(t) \\ x_3(t) \end{bmatrix} \right.
$$
$$
\left. + \begin{bmatrix} u_1(t) & u_2(t) \end{bmatrix} \begin{bmatrix} 0.8 & 0 \\ 0 & 0.4 \end{bmatrix} \begin{bmatrix} u_1(t) \\ u_2(t) \end{bmatrix} \right\} dt \qquad (7.52)
$$

This problem can not be solved using the recursive algorithms in Section 7.2 because the system matrices E and B are not available in the forms shown in Eq. (7.4). Hence, the unified approach in Section 7.3 is applied to compute the optimal control law and the state vector. $\mathbf{u}(t)$ and $\mathbf{x}(t)$ are computed with $m = 12$ in the BPF approach and $m = 6$ in the SLP approach, and the results are shown in Figures 7.3 and 7.4. The J value in each case is given in Table 7.2. It is clear from the figures that the results are consistent and satisfactory.

Table 7.2: Cost function

Method	m	J
BPFs	12	1.40775085814001
SLPs	6	1.40774133946223

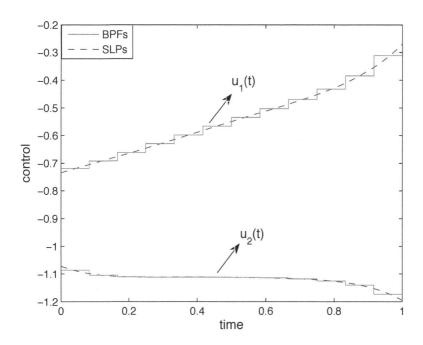

Figure 7.3: SLP and BPF solutions of $\mathbf{u}(t)$.

7.5 Conclusion

In this chapter two recursive algorithms and a unified approach are presented to solve linear quadratic optimal control problem of time-invariant singular systems. The recursive algorithms can be

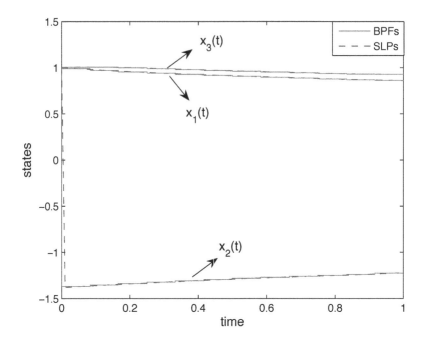

Figure 7.4: SLP and BPF solutions of $\mathbf{x}(t)$.

applied whenever system matrices E, A and B are available, as shown in Eqs. (7.4) and (7.5), and $E_{21} = E_{22} = \bigcirc$. These algorithms are computationally very attractive as they are completely recursive and they involve inversion of matrices of size $2n$ only.

The unified approach involves inversion of a matrix of size mn which becomes larger for large m as the accuracy of computation depends on m value. Though the unified approach is thus not as elegant as the recursive algorithms computationally, it has the merit that it can be applied to any singular system whose matrices need not be in the structured form as defined in Eqs. (7.4) and (7.5). In both the illustrative examples, the results obtained with $m = 6$ SLPs are good. Since BPFs are piecewise constant functions and the actual solution is smooth, one has to choose a large value for m in BPF approach in order to improve upon the accuracy of BPF solution.

Chapter 8

Optimal Control of Time-Delay Systems

Based on using BPFs or SLPs, two unified approaches to compute optimal control law of linear time-delay dynamic systems with quadratic performance index are discussed in this chapter. The governing delay-differential equations of dynamic systems are converted into linear algebraic equations by using the operational matrices of OFs (BPFs and SLPs). Thus, the problem of finding the optimal control law is reduced to the problem of solving algebraic equations obtained via the operational matrices.

First linear time-varying multi-delay dynamic systems are considered and presented as a unified approach to compute the optimal control law and the state variables. Then this algorithm is modified for time-invariant systems and delay-free systems. Next, linear time-varying delay systems with reverse time terms are considered and a unified algorithm for computing the optimal control law is developed. Numerical examples are included to demonstrate the applicability of the approaches.

8.1   Introduction

Time-delay systems are those systems in which time delays exist between the application of input or control to the system and its resulting effect on it. They arise either as a result of inherent delays in the components of the system or as a deliberate introduction of time delay into the system for control purpose. Examples of such systems are electronic systems, mechanical systems, biological systems, environmental systems, metallurgical systems, and chemical systems. A few practical examples [46] are speed control of a steam engine running an electric power generator under varying load conditions, and control of room temperature, a cold rolling mill, spaceship, hydraulic system etc.

As it appears from the literature, extensive work was done on the problem of optimal control of linear continuous-time dynamical systems containing time delays. Palanisamy and Rao [20] appear to be the first to study the optimal control problem via a class of piecewise constant basis functions — WFs. They considered time-invariant systems with a delay in state and a delay in control. In [24] time-varying systems containing a delay in state and a delay in control were considered, and the optimal control problem of such systems was studied via BPFs. Solutions obtained in [20, 24] are piecewise constant. In order to obtain a smooth solution, SCP1s were used [27] to study time-invariant systems with a delay in state only.

In [31] time-varying systems with multiple delays in state and control were studied via SLPs. The problem in [20] was investigated by applying SLPs in [36]. In [48] the problem considered in [24] was solved by approximating the time-delay terms via Pade approximation, and using GOPs. In [58] the problem considered in [27] was studied via SLPs.

In the recent years, people came up with a new idea of defining hybrid functions (with BPFs and any class of polynomial functions) and utilizing the same for studying problems in Systems and Control. The so-called hybrid functions approach was first introduced in [67] to study the optimal control problem of time-varying systems with a time-delay in state only. Subsequently, this approach was extended to time-invariant systems [72] in [27] and to delay systems containing reverse time terms [73]. As a matter of fact, TPBV problems can be transformed into problems that contain reverse time terms in their description. For example [55], the TPBV problem

$$
\begin{aligned}
\dot{y}(t) &= y(t) + y(t - \alpha) + z(t) + a(t) \\
\dot{z}(t) &= y(t) + b(t) \\
y(0) &= y_0, \quad y(t) = \phi_1(t) \ \text{ for } \ -\alpha \le t < 0 \\
z(t_f) &= z_f, \quad z(t) = \phi_2(t) \ \text{ for } \ t_f < t \le t_f + \alpha
\end{aligned}
$$

can be transformed into the following problem containing a reverse time term

$$
\begin{aligned}
\dot{\mathbf{x}}(t) &= E_1\mathbf{x}(t) + E_2\mathbf{x}(t - \alpha) + E_3\mathbf{x}(t_f - t) + \mathbf{u}(t) \\
\mathbf{x}(0) &= \begin{bmatrix} y_0, & z_f \end{bmatrix}^T
\end{aligned}
$$

$$
\mathbf{x}(t) = \begin{bmatrix} \phi_1(t), & \phi_2(t_f - t) \end{bmatrix}^T \ \text{ for } \ -\alpha \le t < 0
$$

where

$$
\mathbf{x}(t) = \begin{bmatrix} y(t), & z(t_f - t) \end{bmatrix}^T, \quad \mathbf{u}(t) = \begin{bmatrix} a(t), & -b(t_f - t) \end{bmatrix}^T,
$$
$$
E_1 = E_2 = \begin{bmatrix} 1 & 0 \\ 0 & 0 \end{bmatrix} \ \text{ and } \ E_3 = \begin{bmatrix} 0 & 1 \\ -1 & 0 \end{bmatrix}.
$$

In [70] the application of a Fourier series to optimal control of linear time invariant systems has been commented on.

Linear Legendre multiwavelets [81] were used to solve the time-varying systems having a delay in state only. Hybrid functions were employed [83] to study the same problem considered in [67]. In [84] general Legendre wavelets were used to solve the optimal control problem of time-varying singular systems having a delay in state only.

Now, the following observations can be made from the above discussions on the historical developments of the optimal control problem of time-delay systems via OFs :

- The BPF approach reported in [24] is restricted to only time-varying systems containing a delay in state and a delay in control. It is not yet available for multi-delay systems.

- The systems considered in [31] are time-varying with multi delays in both state and control. Such systems are studied only via SLPs. The BPF approach is not yet reported.

- The delay systems containing reverse time terms [73] were studied via hybrid functions. This problem is not yet studied using the OFs (SLPs and BPFs).

Therefore, in this chapter, an attempt is made to introduce a unified approach for computing optimal control of linear time-varying systems with multiple delays in both state and control via OFs (BPFs and SLPs). This approach [87, 90, 94] is different from the one in [24, 31] in the following two ways:

In [24, 31], $\dot{\mathbf{x}}(t)$ was expressed in terms of OFs first, the spectrum of $\mathbf{x}(t)$ was expressed in terms of $\dot{\mathbf{x}}(t)$ spectrum, and then

$\mathbf{u}(t)$ was finally calculated. In our approach, the unknown $\mathbf{x}(t)$ is directly expressed in terms of OFs, so that the state feedback control law $\mathbf{u}(t)$ can be expressed in terms of the spectrum of $\mathbf{x}(t)$. Thus our approach is straightforward.

The manner in which the final cost term in performance index is handled in terms of OFs in our approach is different from that in [24, 31].

Next, a unified approach is introduced [91] to compute the optimal control law of linear time-varying systems containing delays and reverse time terms in state and control.

The chapter is organised as follows: The next section deals with optimal control of time-delay/delay free, time-invariant/time-varying systems via BPFs and SLPs. Time-varying systems containing delays and reverse time terms in state and control are considered in Section 8.3. Numerical examples are given to demonstrate the effectiveness of our approaches. The last section concludes the chapter.

8.2 Optimal Control of Multi-Delay Systems

Consider an nth order linear time-varying system with multiple delays

$$\dot{\mathbf{x}}(t) = \sum_{l=0}^{\alpha} A_l(t)\mathbf{x}(t - \tau_l) + \sum_{l=0}^{\beta} B_l(t)\mathbf{u}(t - \theta_l) \quad (8.1)$$

$$\mathbf{x}(t) = \boldsymbol{\zeta}(t) \quad \text{for} \quad t \leq t_0 \quad (8.2)$$

$$\mathbf{u}(t) = \boldsymbol{\nu}(t) \quad \text{for} \quad t \leq t_0 \quad (8.3)$$

where $\mathbf{x}(t)$ is an n dimensional state vector, $\mathbf{u}(t)$ is an r dimensional control vector, $A_l(t)$ and $B_l(t)$ are $n \times n$ and $n \times r$ time-varying matrices, and $\tau_l, \theta_l \geq 0$ are constant time-delays in state and control, respectively.

The objective is to find the control vector $\mathbf{u}(t)$ that minimizes the quadratic cost function

$$J = \frac{1}{2}\mathbf{x}^T(t_f)S\mathbf{x}(t_f) + \frac{1}{2}\int_{t_0}^{t_f} \left[\mathbf{x}^T(t)Q(t)\mathbf{x}(t) + \mathbf{u}^T(t)R(t)\mathbf{u}(t)\right] dt$$

(8.4)

where S and $Q(t)$ are $n \times n$ symmetric positive semidefinite matrices, and $R(t)$ is an $r \times r$ symmetric positive definite matrix.

Integrating Eq. (8.1) once with respect to t, we obtain

$$\mathbf{x}(t) - \mathbf{x}(t_0) = \int_{t_0}^{t} \left[\sum_{l=0}^{\alpha} A_l(\sigma)\mathbf{x}(\sigma - \tau_l) + \sum_{l=0}^{\beta} B_l(\sigma)\mathbf{u}(\sigma - \theta_l)\right] d\sigma$$

(8.5)

We express $\mathbf{x}(t)$, $\mathbf{u}(t)$, $\mathbf{x}(t_0)$, $\mathbf{x}(t - \tau_l)$, $\mathbf{u}(t - \theta_l)$, $A_l(t)$ and $B_l(t)$ in terms of OFs $\{\phi_i(t)\}$ (BPFs or SLPs) as follows:

$$\mathbf{x}(t) \approx \sum_{i=0}^{m-1} \mathbf{x}_i\phi_i(t) = X\phi(t) \tag{8.6}$$

$$\mathbf{u}(t) \approx \sum_{i=0}^{m-1} \mathbf{u}_i\phi_i(t) = U\phi(t) \tag{8.7}$$

$$\mathbf{x}(t_0) = V\phi(t) \tag{8.8}$$

$$\mathbf{x}(t - \tau_l) \approx X^*(\tau_l)\phi(t) \tag{8.9}$$

$$\mathbf{u}(t - \theta_l) \approx U^*(\theta_l)\phi(t) \tag{8.10}$$

$$A_l(t) \approx \sum_{i=0}^{m-1} A_{li}\phi_i(t) \tag{8.11}$$

$$B_l(t) \approx \sum_{i=0}^{m-1} B_{li}\phi_i(t) \tag{8.12}$$

$$A_l(t)\mathbf{x}(t - \tau_l) \approx \sum_{i=0}^{m-1} \mathbf{y}_i^\star(\tau_l)\phi_i(t) = Y^\star(\tau_l)\boldsymbol{\phi}(t) \quad (8.13)$$

$$B_l(t)\mathbf{u}(t - \theta_l) \approx \sum_{i=0}^{m-1} \mathbf{z}_i^\star(\theta_l)\phi_i(t) = Z^\star(\theta_l)\boldsymbol{\phi}(t) \quad (8.14)$$

where $\boldsymbol{\phi}(t)$ is, given in Eq. (7.12), an m dimensional OFs vector, i.e. $\boldsymbol{\phi}(t)$ is $\mathbf{B}(t)$ or $\mathbf{L}(t)$. Substituting Eqs. (8.6), (8.8), (8.12) and (8.14) into Eq. (8.5), and using forward integration operational property in Eq. (2.10)/(2.50) of OFs yield

$$X - V = \left[\sum_{l=0}^{\alpha} Y^\star(\tau_l) + \sum_{l=0}^{\beta} Z^\star(\theta_l) \right] H_f$$

$$\Rightarrow \quad \hat{\mathbf{x}} = \hat{\mathbf{v}} + \left(H_f^T \otimes I_n \right) \left[\sum_{l=0}^{\alpha} \hat{\mathbf{y}}^\star(\tau_l) + \sum_{l=0}^{\beta} \hat{\mathbf{z}}^\star(\theta_l) \right] \quad (8.15)$$

where $\otimes$ is the Kronecker product of matrices [12],

$$\hat{\mathbf{x}} = \begin{bmatrix} \mathbf{x}_0 \\ \mathbf{x}_1 \\ \vdots \\ \mathbf{x}_{m-1} \end{bmatrix} ; \ \hat{\mathbf{v}} = \begin{bmatrix} \mathbf{v}_0 \\ \mathbf{v}_1 \\ \vdots \\ \mathbf{v}_{m-1} \end{bmatrix} ; \ \hat{\mathbf{y}}^\star(\tau_l) = \begin{bmatrix} \mathbf{y}_0^\star(\tau_l) \\ \mathbf{y}_1^\star(\tau_l) \\ \vdots \\ \mathbf{y}_{m-1}^\star(\tau_l) \end{bmatrix} ;$$

$$\hat{\mathbf{z}}^\star(\theta_l) = \begin{bmatrix} \mathbf{z}_0^\star(\theta_l) \\ \mathbf{z}_1^\star(\theta_l) \\ \vdots \\ \mathbf{z}_{m-1}^\star(\theta_l) \end{bmatrix} \quad (8.16)$$

It is possible to write $\hat{\mathbf{y}}^\star(\tau_l)$ and $\hat{\mathbf{z}}^\star(\theta_l)$ as

$$\hat{\mathbf{y}}^\star(\tau_l) = \hat{A}_l \hat{\mathbf{x}}^\star(\tau_l) \quad \text{and} \quad \hat{\mathbf{z}}^\star(\theta_l) = \hat{B}_l \hat{\mathbf{u}}^\star(\theta_l) \quad (8.17)$$

where

$$\hat{\mathbf{x}}^\star(\tau_l) = \begin{bmatrix} \mathbf{x}_0^\star(\tau_l) \\ \mathbf{x}_1^\star(\tau_l) \\ \vdots \\ \mathbf{x}_{m-1}^\star(\tau_l) \end{bmatrix} ; \ \hat{\mathbf{u}}^\star(\theta_l) = \begin{bmatrix} \mathbf{u}_0^\star(\theta_l) \\ \mathbf{u}_1^\star(\theta_l) \\ \vdots \\ \mathbf{u}_{m-1}^\star(\theta_l) \end{bmatrix} \quad (8.18)$$

and $\hat{A}_l$ and $\hat{B}_l$ are defined in the following subsections. Eq. (8.15) can be rewritten in the form

$$\hat{\mathbf{x}} = M\hat{\mathbf{u}} + \hat{\mathbf{w}} \tag{8.19}$$

where $\hat{\mathbf{u}}$ is similar to $\hat{\mathbf{x}}$ in Eq. (8.16), and M and $\hat{\mathbf{w}}$ are defined in the following subsections.

Now the final cost term in Eq. (8.4) can be written as

$$\mathbf{x}^T(t_f)S\mathbf{x}(t_f) = [\mathbf{x}(t_0) + \Omega\,\hat{\mathbf{x}}]^T S [\mathbf{x}(t_0) + \Omega\,\hat{\mathbf{x}}]$$

where

$$\Omega = 2\left(\mathbf{b}^T \otimes I_n\right) \tag{8.20}$$

and $\mathbf{b}$ is defined in the following subsections. Expressing $Q(t)$ and $R(t)$ in terms of OFs, we have

$$Q(t) \approx \sum_{i=0}^{m-1} Q_i \phi_i(t) \tag{8.21}$$

$$R(t) \approx \sum_{i=0}^{m-1} R_i \phi_i(t) \tag{8.22}$$

and

$$\int_{t_0}^{t_f} \left[\mathbf{x}^T(t)Q(t)\mathbf{x}(t) + \mathbf{u}^T(t)R(t)\mathbf{u}(t)\right] dt \simeq \hat{\mathbf{x}}^T \hat{Q}\hat{\mathbf{x}} + \hat{\mathbf{u}}^T \hat{R}\hat{\mathbf{u}}$$

where $\hat{Q}$ and $\hat{R}$ are defined in the following subsections. So Eq. (8.4) becomes

$$J = \frac{1}{2}[\mathbf{x}(t_0) + \Omega\,\hat{\mathbf{x}}]^T S [\mathbf{x}(t_0) + \Omega\,\hat{\mathbf{x}}] + \frac{1}{2}\left(\hat{\mathbf{x}}^T \hat{Q}\hat{\mathbf{x}} + \hat{\mathbf{u}}^T \hat{R}\hat{\mathbf{u}}\right) \tag{8.23}$$

Substituting Eq. (8.19) into Eq. (8.23) and setting the optimization condition given in Eq. (7.41) yield the optimal control law

[87, 90, 94]

$$\hat{\mathbf{u}} = -\left[M^T\left(\Omega^T S\Omega + \hat{Q}\right)M + \hat{R}\right]^{-1}$$
$$\times \left[M^T\Omega^T S\mathbf{x}(t_0) + M^T\left(\Omega^T S\Omega + \hat{Q}\right)\hat{\mathbf{w}}\right] \quad (8.24)$$

8.2.1 Using BPFs

In Eq. (8.16), we have [87, 90]

$$\hat{\mathbf{v}} = \left[\ \mathbf{x}^T(t_0),\ \mathbf{x}^T(t_0),\ \ldots,\ \mathbf{x}^T(t_0)\ \right]^T \quad (8.25)$$

In Eq. (8.17)

$$\hat{A}_l = \mathrm{diag}\left[\ A_{l0},\ A_{l1},\ \ldots,\ A_{l,m-1}\ \right] \quad (8.26)$$
$$\hat{B}_l = \mathrm{diag}\left[\ B_{l0},\ B_{l1},\ \ldots,\ B_{l,m-1}\ \right] \quad (8.27)$$
$$\hat{\mathbf{x}}^\star(\tau_l) = E\left(n,\ \mu_l\right)\hat{\boldsymbol{\zeta}}(\tau_l) + D\left(n,\ \mu_l\right)\hat{\mathbf{x}} \quad (8.28)$$
$$\hat{\mathbf{u}}^\star(\theta_l) = E\left(r,\ \delta_l\right)\hat{\boldsymbol{\nu}}(\theta_l) + D\left(r,\ \delta_l\right)\hat{\mathbf{u}} \quad (8.29)$$

where

$$\tau_l = \mu_l T \quad \text{and} \quad \theta_l = \delta_l T \quad (8.30)$$

μ_l and δ_l represent number of BPFs on $t_0 \le t \le t_0 + \tau_l$ and $t_0 \le t \le t_0 + \theta_l$ respectively,

$$\hat{\boldsymbol{\zeta}}(\tau_l) = \begin{bmatrix} \boldsymbol{\zeta}_0(\tau_l) \\ \boldsymbol{\zeta}_1(\tau_l) \\ \vdots \\ \boldsymbol{\zeta}_{\mu_l-1}(\tau_l) \end{bmatrix} ; \quad \hat{\boldsymbol{\nu}}(\theta_l) = \begin{bmatrix} \boldsymbol{\nu}_0(\theta_l) \\ \boldsymbol{\nu}_1(\theta_l) \\ \vdots \\ \boldsymbol{\nu}_{\delta_l-1}(\theta_l) \end{bmatrix} \quad (8.31)$$

$$\boldsymbol{\zeta}_i(\tau_l) = \frac{1}{T}\int_{t_0+iT}^{t_0+(i+1)T} \boldsymbol{\zeta}(t-\tau_l)dt \quad \text{for } i = 0,1,2,\ldots,\mu_l-1$$
$$(8.32)$$

and

$$\boldsymbol{\nu}_i(\theta_l) = \frac{1}{T} \int_{t_0+iT}^{t_0+(i+1)T} \boldsymbol{\nu}(t-\theta_l)dt \quad \text{for } i = 0, 1, 2, \ldots, \delta_l - 1 \tag{8.33}$$

In Eq. (8.19)

$$M = N^{-1} \left(H^T \otimes I_n\right) \sum_{l=0}^{\beta} \hat{B}_l D\left(r, \ \delta_l\right) \tag{8.34}$$

$$\hat{\mathbf{w}} = N^{-1} \left\{ \hat{\mathbf{v}} + \left(H^T \otimes I_n\right) \left[\sum_{l=0}^{\alpha} \hat{A}_l E\left(n, \mu_l\right) \hat{\boldsymbol{\zeta}}(\tau_l) \right. \right.$$
$$\left. \left. + \sum_{l=0}^{\beta} \hat{B}_l E\left(r, \delta_l\right) \hat{\boldsymbol{\nu}}(\theta_l) \right] \right\} \tag{8.35}$$

where

$$N = I_{mn} - \left(H^T \otimes I_n\right) \sum_{l=0}^{\alpha} \hat{A}_l D\left(n, \ \mu_l\right) \tag{8.36}$$

In Eq. (8.20)

$$\mathbf{b} = \begin{bmatrix} -1, & 1, & -1, & 1, & \ldots, & -1, & 1 \end{bmatrix}^T \tag{8.37}$$

an m dimensional vector. In Eq. (8.23)

$$\hat{Q} = T \times \text{diag} \begin{bmatrix} Q_0, & Q_1, & \ldots, & Q_{m-1} \end{bmatrix} \tag{8.38}$$

and

$$\hat{R} = T \times \text{diag} \begin{bmatrix} R_0, & R_1, & \ldots, & R_{m-1} \end{bmatrix} \tag{8.39}$$

8.2.2 Using SLPs

In Eq. (8.16), we have [87, 94]

$$\hat{\mathbf{v}} = \left[\mathbf{x}^T(t_0), \ \mathbf{0}^T, \ \ldots, \ \mathbf{0}^T \right]^T \tag{8.40}$$

$$\hat{\mathbf{y}}_i^{\star}(\tau_l) = \frac{(2i+1)}{(t_f - t_0)} \sum_{j=0}^{m-1} \sum_{k=0}^{m-1} \pi_{ijk} A_{lj} \, \hat{\mathbf{x}}_k^{\star}(\tau_l) \tag{8.41}$$

$$\hat{\mathbf{z}}_i^{\star}(\theta_l) = \frac{(2i+1)}{(t_f - t_0)} \sum_{j=0}^{m-1} \sum_{k=0}^{m-1} \pi_{ijk} B_{lj} \, \hat{\mathbf{u}}_k^{\star}(\theta_l) \tag{8.42}$$

In Eq. (8.17)

$$\hat{A}_l = \frac{1}{(t_f - t_0)} \begin{bmatrix} \sum\limits_{j=0}^{m-1} \pi_{0j0} A_{lj} & \cdots & \sum\limits_{j=0}^{m-1} \pi_{0j,m-1} A_{lj} \\ 3 \sum\limits_{j=0}^{m-1} \pi_{1j0} A_{lj} & \cdots & 3 \sum\limits_{j=0}^{m-1} \pi_{1j,m-1} A_{lj} \\ \vdots & & \vdots \\ (2m-1) \sum\limits_{j=0}^{m-1} \pi_{m-1,j0} A_{lj} & \cdots & (2m-1) \sum\limits_{j=0}^{m-1} \pi_{m-1,j,m-1} A_{lj} \end{bmatrix} \tag{8.43}$$

$$\hat{B}_l = \frac{1}{(t_f - t_0)} \begin{bmatrix} \sum\limits_{j=0}^{m-1} \pi_{0j0} B_{lj} & \cdots & \sum\limits_{j=0}^{m-1} \pi_{0j,m-1} B_{lj} \\ 3 \sum\limits_{j=0}^{m-1} \pi_{1j0} B_{lj} & \cdots & 3 \sum\limits_{j=0}^{m-1} \pi_{1j,m-1} B_{lj} \\ \vdots & & \vdots \\ (2m-1) \sum\limits_{j=0}^{m-1} \pi_{m-1,j0} B_{lj} & \cdots & (2m-1) \sum\limits_{j=0}^{m-1} \pi_{m-1,j,m-1} B_{lj} \end{bmatrix} \tag{8.44}$$

$$\hat{\mathbf{x}}^{\star}(\tau_l) = \hat{\boldsymbol{\zeta}}(\tau_l) + (D(\tau_l) \otimes I_n) \, \hat{\mathbf{x}} \tag{8.45}$$

$$\hat{\mathbf{u}}^{\star}(\theta_l) = \hat{\boldsymbol{\nu}}(\theta_l) + (D(\theta_l) \otimes I_r) \, \hat{\mathbf{u}} \tag{8.46}$$

where

$$\hat{\boldsymbol{\zeta}}(\tau_l) = \begin{bmatrix} \boldsymbol{\zeta}_0(\tau_l) \\ \boldsymbol{\zeta}_1(\tau_l) \\ \vdots \\ \boldsymbol{\zeta}_{m-1}(\tau_l) \end{bmatrix} ; \quad \hat{\boldsymbol{\nu}}(\theta_l) = \begin{bmatrix} \boldsymbol{\nu}_0(\theta_l) \\ \boldsymbol{\nu}_1(\theta_l) \\ \vdots \\ \boldsymbol{\nu}_{m-1}(\theta_l) \end{bmatrix} \tag{8.47}$$

$$\boldsymbol{\zeta}_i(\tau_l) = \frac{(2i+1)}{(t_f - t_0)} \int_{t_0}^{t_0+\tau_l} \boldsymbol{\zeta}(t-\tau_l) L_i(t) dt \tag{8.48}$$

$$\boldsymbol{\nu}_i(\theta_l) = \frac{(2i+1)}{(t_f - t_0)} \int_{t_0}^{t_0+\theta_l} \boldsymbol{\nu}(t-\theta_l) L_i(t) dt \tag{8.49}$$

for $i = 0, 1, 2, \ldots, m-1$. In Eq. (8.19)

$$M = N^{-1} \left(H^T \otimes I_n \right) \sum_{l=0}^{\beta} \hat{B}_l \left(D(\theta_l) \otimes I_r \right) \tag{8.50}$$

$$\hat{\mathbf{w}} = N^{-1} \left\{ \hat{\mathbf{v}} + \left(H^T \otimes I_n \right) \left[\sum_{l=0}^{\alpha} \hat{A}_l \hat{\boldsymbol{\zeta}}(\tau_l) + \sum_{l=0}^{\beta} \hat{B}_l \hat{\boldsymbol{\nu}}(\theta_l) \right] \right\} \tag{8.51}$$

where

$$N = I_{mn} - \left(H^T \otimes I_n \right) \sum_{l=0}^{\alpha} \hat{A}_l \left(D(\tau_l) \otimes I_n \right) \tag{8.52}$$

In Eq. (8.20)

$$\mathbf{b} = \begin{bmatrix} 0, & 1, & 0, & 1, & \ldots, & 0, & 1 \end{bmatrix}^T \tag{8.53}$$

an m dimensional vector. In Eq. (8.23)

$$\hat{Q} = \begin{bmatrix} \sum_{j=0}^{m-1} \pi_{0j0}Q_j & \sum_{j=0}^{m-1} \pi_{0j1}Q_j & \cdots & \sum_{j=0}^{m-1} \pi_{0j,m-1}Q_j \\ \sum_{j=0}^{m-1} \pi_{1j0}Q_j & \sum_{j=0}^{m-1} \pi_{1j1}Q_j & \cdots & \sum_{j=0}^{m-1} \pi_{1j,m-1}Q_j \\ \vdots & \vdots & & \vdots \\ \sum_{j=0}^{m-1} \pi_{m-1,j0}Q_j & \sum_{j=0}^{m-1} \pi_{m-1,j1}Q_j & \cdots & \sum_{j=0}^{m-1} \pi_{m-1,j,m-1}Q_j \end{bmatrix}$$

$$\text{(8.54)}$$

$$\hat{R} = \begin{bmatrix} \sum_{j=0}^{m-1} \pi_{0j0}R_j & \sum_{j=0}^{m-1} \pi_{0j1}R_j & \cdots & \sum_{j=0}^{m-1} \pi_{0j,m-1}R_j \\ \sum_{j=0}^{m-1} \pi_{1j0}R_j & \sum_{j=0}^{m-1} \pi_{1j1}R_j & \cdots & \sum_{j=0}^{m-1} \pi_{1j,m-1}R_j \\ \vdots & \vdots & & \vdots \\ \sum_{j=0}^{m-1} \pi_{m-1,j0}R_j & \sum_{j=0}^{m-1} \pi_{m-1,j1}R_j & \cdots & \sum_{j=0}^{m-1} \pi_{m-1,j,m-1}R_j \end{bmatrix}$$

$$\text{(8.55)}$$

8.2.3   Time-invariant systems

The algorithms discussed above are applicable to time-varying systems. With the following changes [87, 90, 94], they are also applicable to time-invariant systems as the time-varying matrices $A_l(t)$, $B_l(t)$, $Q(t)$ and $R(t)$ are constant for such systems. Therefore,

$$\hat{A}_l = I_m \otimes A_l, \quad \hat{B}_l = I_m \otimes B_l \tag{8.56}$$

Via BPFs

$$\hat{Q} = T\,(I_m \otimes Q), \quad \hat{R} = T\,(I_m \otimes R) \tag{8.57}$$

Via SLPs

$$\hat{Q} = \nabla \otimes Q, \quad \hat{R} = \nabla \otimes R \tag{8.58}$$

where

$$\nabla = (t_f - t_0)\,\text{diag}\left[\, 1, \tfrac{1}{3}, \cdots, \tfrac{1}{2m-1} \,\right] \tag{8.59}$$

8.2.4 Delay free systems

In this case $\alpha = \beta = 0$, $\tau_l = \theta_l = 0$, and $\mathbf{x}(t) = \mathbf{x}(t_0)$ at $t = t_0$. Therefore [87, 90, 94]

$$\hat{\zeta}(\tau_l) = \hat{\nu}(\theta_l) = \mathbf{0} \tag{8.60}$$

Via BPFs

$$D(n, \mu_l) = I_{mn} \quad \text{and} \quad D(r, \delta_l) = I_{mr} \tag{8.61}$$

Via SLPs

$$D(\tau_l) = D(\theta_l) = I_m \tag{8.62}$$

8.2.5 Illustrative examples

In this subsection five different examples are considered. The first three are for regular time-delay systems and the last two examples are for singular time-delay systems.

Example 8.1

Consider the time-invariant system [31] with a delay in state

$$\dot{x}(t) = x(t-1) + u(t)$$
$$x(t) = 1 \quad \text{for} \quad -1 \le t \le 0$$

with the cost function

$$J = \frac{1}{2}\left[10^5 x^2(2) + \int_0^2 u^2(t)dt\right]$$

The exact solution is given by

$$u(t) = \begin{cases} -2.1 + 1.05t & \text{for } 0 \le t \le 1 \\ -1.05 & \text{for } 1 \le t \le 2 \end{cases}$$

$$x(t) = \begin{cases} 1 - 1.1t + 0.525t^2 & \text{for } 0 \le t \le 1 \\ -0.25 + 1.575t - 1.075t^2 + 0.175t^3 & \text{for } 1 \le t \le 2 \end{cases}$$

Figure 8.1 shows $u(t)$ and $x(t)$ obtained via the BPF and SLP approaches with $m = 8$, and the analytical method. The value of J is shown in Table 8.1. The results match well with the exact results in each case.

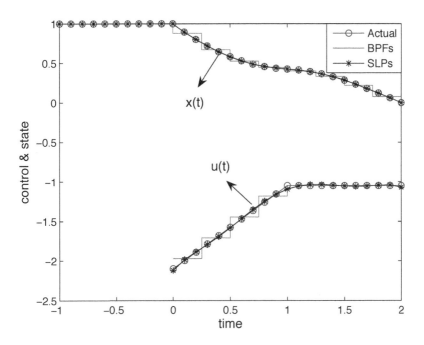

Figure 8.1: Exact, BPF and SLP solutions of $u(t)$ and $x(t)$ variables.

Table 8.1: Cost function

Method	J
Exact	1.8375
BPF	1.8404
SLP	1.8379

Example 8.2

Consider the following time-invariant system [67] with a delay in state and a delay in control

$$\dot{x}(t) = -x(t) + x\left(t - \frac{1}{3}\right) + u(t) - \frac{1}{2}u\left(t - \frac{2}{3}\right)$$

$$x(t) = 1 \quad \text{for} \quad -\frac{1}{3} \leq t \leq 0$$

$$u(t) = 0 \quad \text{for} \quad -\frac{2}{3} \leq t \leq 0$$

with the performance index

$$J = \frac{1}{2} \int_0^1 \left[x^2(t) + \frac{1}{2} u^2(t) \right] dt$$

This problem is solved using BPFs and SLPs with $m = 12$. The values of $u(t)$ and $x(t)$ are calculated and shown in Fig. 8.2. For comparison purpose, the results obtained by hybrid functions approach (# of BPFs = 3 and # of SLPs = 4) [67] are also reproduced in the same figure. The J values obtained by the BPF and SLP approaches, and the hybrid functions approach [67] are presented in Table 8.2. From the figure and the table it is clear that the results match each other.

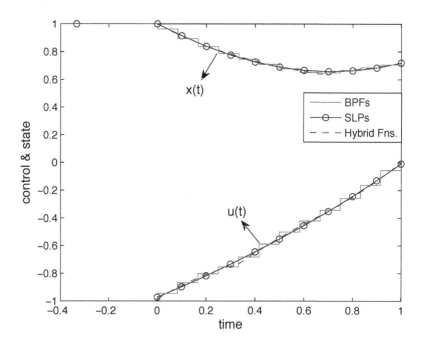

Figure 8.2: BPF, SLP and hybrid function [67] solutions of $u(t)$ and $x(t)$.

Table 8.2: Cost function

Method	J
BPF	0.3731831
SLP	0.3731102
Hybrid [67]	0.3731129

Example 8.3

Consider the time-varying multi-delay system [31]

$$
\begin{bmatrix} \dot{x}_1(t) \\ \dot{x}_2(t) \end{bmatrix} = \begin{bmatrix} 0 & 1 \\ t & 0 \end{bmatrix} \begin{bmatrix} x_1(t) \\ x_2(t) \end{bmatrix} + \begin{bmatrix} 0 & 0 \\ 0 & 1 \end{bmatrix} \begin{bmatrix} x_1(t-0.8) \\ x_2(t-0.8) \end{bmatrix}
$$

$$
+ \begin{bmatrix} 1 & 0 \\ 2 & 0 \end{bmatrix} \begin{bmatrix} x_1(t-1) \\ x_2(t-1) \end{bmatrix} + \begin{bmatrix} 0 \\ 1 \end{bmatrix} u(t)
$$

$$
+ \begin{bmatrix} 0 \\ -1 \end{bmatrix} u(t-0.5)
$$

$$
\begin{bmatrix} x_1(t) \\ x_2(t) \end{bmatrix} = \begin{bmatrix} 1 \\ 1 \end{bmatrix} \quad \text{for} \quad -1 \le t \le 0
$$

$$
u(t) = 5(t+1) \quad \text{for} \quad -0.5 \le t \le 0
$$

with the cost function

$$
J = \frac{1}{2} \begin{bmatrix} x_1(3) & x_2(3) \end{bmatrix} \begin{bmatrix} 1 & 0 \\ 0 & 2 \end{bmatrix} \begin{bmatrix} x_1(3) \\ x_2(3) \end{bmatrix}
$$

$$
+ \frac{1}{2} \int_0^3 \left\{ \begin{bmatrix} x_1(t) & x_2(t) \end{bmatrix} \begin{bmatrix} 2 & 1 \\ 1 & 1 \end{bmatrix} \begin{bmatrix} x_1(t) \\ x_2(t) \end{bmatrix} + \frac{1}{2+t} u^2(t) \right\} dt
$$

The optimal control law $u(t)$ and the state variables $x_1(t)$ and $x_2(t)$ are computed with $m = 30$ in the BPF approach and $m = 12$ in the SLP approach, and the results are shown in Fig. 8.3. The value of J in each case is given in Table 8.3.

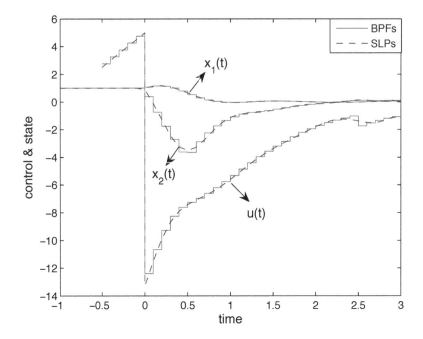

Figure 8.3: BPF and SLP solutions of $u(t)$ and $\mathbf{x}(t)$ variables.

Table 8.3: Cost function

Method	m	J
BPF	30	19.6414
SLP	12	19.7079

Example 8.4

Consider the singular time-delay system [84]

$$\begin{bmatrix} 2 & 1 \\ 6 & 3 \end{bmatrix} \begin{bmatrix} \dot{x}_1(t) \\ \dot{x}_2(t) \end{bmatrix} = \begin{bmatrix} 1 & 2 \\ 3 & 0 \end{bmatrix} \begin{bmatrix} x_1(t) \\ x_2(t) \end{bmatrix} + \begin{bmatrix} 5 & 0 \\ 4 & 1 \end{bmatrix} \begin{bmatrix} x_1\left(t - \frac{1}{3}\right) \\ x_2\left(t - \frac{1}{3}\right) \end{bmatrix}$$

$$+ \begin{bmatrix} 1 \\ 1 \end{bmatrix} u(t), \quad t \in [0, 1]$$

$$\begin{bmatrix} x_1(t) \\ x_2(t) \end{bmatrix} = \begin{bmatrix} 1 \\ 1 \end{bmatrix}, \quad t \in \left[-\frac{1}{3}, 0\right]$$

with the cost function

$$J = \frac{1}{2} \int_0^1 \left\{ \begin{bmatrix} x_1(t) & x_2(t) \end{bmatrix} \begin{bmatrix} 1 & 1 \\ 1 & 1 \end{bmatrix} \begin{bmatrix} x_1(t) \\ x_2(t) \end{bmatrix} + u^2(t) \right\} dt$$

To solve the above singular time-delay problem using BPFs and SLPs, the approach given in Sec. 8.2 is modified as follows:

- The optimal control law:

$$\hat{\mathbf{u}} = -\left(M^T \hat{Q} M + \hat{R} \right)^{-1} M^T \hat{Q} \hat{\mathbf{w}} \qquad (8.63)$$

where

$$M = N \left(H_f^T \otimes B \right) \qquad (8.64)$$
$$\hat{\mathbf{w}} = N \left[(I_m \otimes E) \hat{\mathbf{x}}_0 + \left(H_f^T \otimes C \right) E\left(n, \ \mu \right) \hat{\zeta}(\tau) \right]$$
$$\text{for BPFs} \qquad (8.65)$$
$$= N \left[(I_m \otimes E) \hat{\mathbf{x}}_0 + \left(H_f^T \otimes C \right) \hat{\zeta}(\tau) \right] \text{ for SLPs} \quad (8.66)$$
$$N = \left[(I_m \otimes E) - \left(H_f^T \otimes A \right) - \left(H_f^T \otimes C \right) D\left(n, \ \mu \right) \right]^{-1}$$
$$\text{for BPFs} \qquad (8.67)$$
$$= \left[(I_m \otimes E) - \left(H_f^T \otimes A \right) - \left(H_f^T \otimes C \right) (D(\tau) \otimes I_n) \right]^{-1}$$
$$\text{for SLPs} \qquad (8.68)$$

- For BPF approach $E\left(n, \ \mu \right)$, $D\left(n, \ \mu \right)$, $\hat{\zeta}(\tau)$ are given in Sec. 2.1.4, and $\hat{Q}$, $\hat{R}$ are given in Eq. (8.57).

- For SLP approach $D(\tau)$ and $\hat{\zeta}(\tau)$ are given in Sec. 2.2.4, and $\hat{Q}$, $\hat{R}$ are given in Eqs. (8.58) and (8.59).

Now with the above modified equations the values of $u(t)$ and $\mathbf{x}(t)$ are calculated with $m = 6$ BPFs and $m = 12$ SLPs, and shown in Figures 8.4 and 8.5 respectively. For comparison, the values of

$u(t)$ and $\mathbf{x}(t)$, obtained via general Legendre wavelets [84], are also shown in the same figures. The J values obtained by the BPF and SLP approaches and by the general Legendre wavelets [84] are presented in Table 8.4. From the figures and the table it is clear that our approach gives better and more consistent results in comparison with the results reported in [84].

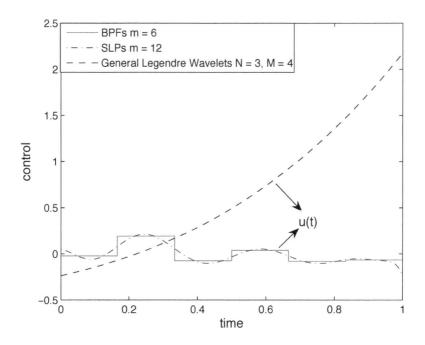

Figure 8.4: BPF, SLP and general Legendre wavelets [84] solutions of $u(t)$.

Table 8.4: Cost function

Method	m	J
BPF	6	0.25512695180308
SLP	12	0.24890228943615
General Legendre wavelets [84]	$N = 3, M = 4$	0.26098867936334

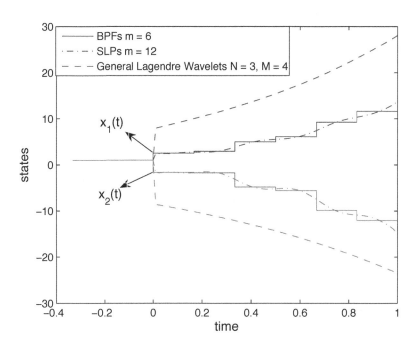

Figure 8.5: BPF, SLP and general Legendre wavelets [84] solutions of $\mathbf{x}(t)$.

Example 8.5

Consider the 3rd order singular time-delay system [84]

$$
\begin{bmatrix} 2.3425 & 0 & 1.4644 \\ 0 & 0 & 0 \\ 3.8654 & 0 & 1.4453 \end{bmatrix} \begin{bmatrix} \dot{x}_1(t) \\ \dot{x}_2(t) \\ \dot{x}_3(t) \end{bmatrix} = \begin{bmatrix} 3.3657 & 0.3563 & -1.8445 \\ 0.5823 & 1.4763 & 4.7343 \\ -1.7623 & 0.2355 & 2.7433 \end{bmatrix} \begin{bmatrix} x_1(t) \\ x_2(t) \\ x_3(t) \end{bmatrix}
$$

$$
+ \begin{bmatrix} 3.3567 & 2.4587 & 2.2157 \\ 1.8743 & 3.5622 & 1.3622 \\ 1.7623 & 1.2243 & 3.5342 \end{bmatrix} \begin{bmatrix} x_1\left(t - \frac{1}{4}\right) \\ x_2\left(t - \frac{1}{4}\right) \\ x_3\left(t - \frac{1}{4}\right) \end{bmatrix}
$$

$$
+ \begin{bmatrix} 1.4874 & 0.4207 \\ 1.0442 & 2.3464 \\ 1.5642 & 0.4768 \end{bmatrix} \begin{bmatrix} u_1(t) \\ u_2(t) \end{bmatrix}, \quad t \in [0, 1]
$$

$$
\begin{bmatrix} x_1(t) \\ x_2(t) \\ x_3(t) \end{bmatrix} = \begin{bmatrix} 1 \\ 1 \\ 1 \end{bmatrix}, \quad t \in \left[-\frac{1}{4}, 0\right]
$$

with the cost function

$$J = \frac{1}{2} \int_0^1 \left\{ \begin{bmatrix} x_1(t) & x_2(t) & x_3(t) \end{bmatrix} \begin{bmatrix} 0.5473 & 1.4369 & 0.5634 \\ 1.4369 & 1.4465 & 0.2467 \\ 0.5634 & 0.2467 & 0.4567 \end{bmatrix} \begin{bmatrix} x_1(t) \\ x_2(t) \\ x_3(t) \end{bmatrix} \right.$$
$$\left. + \begin{bmatrix} u_1(t) & u_2(t) \end{bmatrix} \begin{bmatrix} 0.8 & 0 \\ 0 & 0.4 \end{bmatrix} \begin{bmatrix} u_1(t) \\ u_2(t) \end{bmatrix} \right\} dt$$

Using Eqs. (8.63)–(8.68) the optimal control law $\mathbf{u}(t)$ and the state vector $\mathbf{x}(t)$ are computed with $m = 8$ in the BPF approach and $m = 12$ in the SLP approach, and the results are shown in Figures 8.6 and 8.7. The value of J in each case is given in Table 8.5. For comparison purpose, the J value obtained by general Legendre wavelets method is also given in the same table. Once again it is observed that the results obtained by general Legendre wavelets method [84] are inferior.

Table 8.5: Cost function

Method	m	J
BPF	8	1.58750201119855
SLP	12	1.54006966446442
General Legendre wavelets [84]	$N = 12, M = 4$	1.59386030312733

8.3 Optimal Control of Delay Systems with Reverse Time Terms

Consider the system described by

$$\dot{\mathbf{x}}(t) = A(t)\mathbf{x}(t) + B(t)\mathbf{u}(t) + C(t)\mathbf{x}(t - \tau) + K(t)\mathbf{u}(t - \theta)$$
$$+ G(t)\mathbf{x}(t_0 + t_f - t) + P(t)\mathbf{u}(t_0 + t_f - t) \quad (8.69)$$
$$\mathbf{x}(t) = \zeta(t) \quad \text{for} \quad t \le t_0 \quad (8.70)$$
$$\mathbf{u}(t) = \nu(t) \quad \text{for} \quad t \le t_0 \quad (8.71)$$

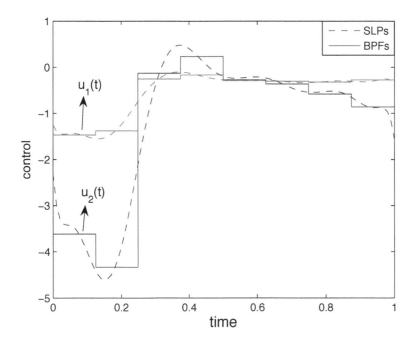

Figure 8.6: BPF and SLP solutions of $u(t)$.

where $\mathbf{x}(t)$ is an n dimensional state vector, $\mathbf{u}(t)$ is an r dimensional control vector, $A(t)$, $B(t)$, $C(t)$, $K(t)$, $G(t)$ and $P(t)$ are time-varying matrices of appropriate dimensions, and τ, $\theta \geq 0$ are constant time-delays in state and control, respectively.

The aim is to find the control law $\mathbf{u}(t)$ which minimizes the quadratic cost function

$$J = \frac{1}{2} \int_{t_0}^{t_f} \left[\mathbf{x}^T(t)Q(t)\mathbf{x}(t) + \mathbf{u}^T(t)R(t)\mathbf{u}(t) \right] dt \quad (8.72)$$

where $Q(t)$ and $R(t)$ are symmetric positive semidefinite and symmetric positive definite matrices, respectively.

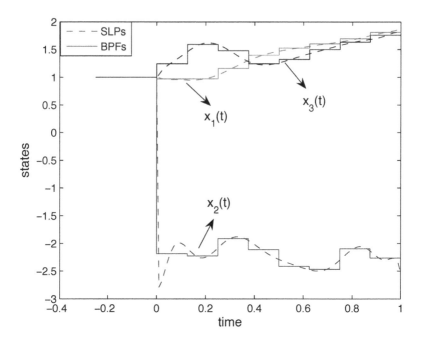

Figure 8.7: BPF and SLP solutions of $\mathbf{x}(t)$.

Integrating Eq. (8.69) with respect to t, we obtain

$$\mathbf{x}(t) - \mathbf{x}(t_0) = \int_{t_0}^{t} [A(\sigma)\mathbf{x}(\sigma) + B(\sigma)\mathbf{u}(\sigma) + C(\sigma)\mathbf{x}(\sigma - \tau)$$

$$+ K(\sigma)\mathbf{u}(\sigma - \theta) + G(\sigma)\mathbf{x}(t_0 + t_f - \sigma)$$

$$+ P(\sigma)\mathbf{u}(t_0 + t_f - \sigma)] \, d\sigma \qquad (8.73)$$

Now we express all the terms in Eq. (8.73) in terms of OFs $\{\phi_i(t)\}$ (BPFs or SLPs). Expressions for $\mathbf{x}(t)$, $\mathbf{u}(t)$, $\mathbf{x}(t_0)$ are given in Eqs. (8.6)–(8.8), $\mathbf{x}(t - \tau)$ and $\mathbf{u}(t - \theta)$ are similar to Eqs. (8.9) and

(8.10), and $A(t)$, $B(t)$, $C(t)$, $K(t)$, $G(t)$ and $P(t)$ are similar to Eq. (8.11)/(8.12).

$$\mathbf{x}(t_0 + t_f - t) \approx \check{X}\boldsymbol{\phi}(t) = X\check{I}\boldsymbol{\phi}(t) \tag{8.74}$$

$$\mathbf{u}(t_0 + t_f - t) \approx \check{U}\boldsymbol{\phi}(t) = U\check{I}\boldsymbol{\phi}(t) \tag{8.75}$$

$$A(t)\mathbf{x}(t) = \tilde{\mathbf{a}}(t) \approx \sum_{i=0}^{m-1} \tilde{\mathbf{a}}_i\phi_i(t) = \tilde{A}\boldsymbol{\phi}(t) \tag{8.76}$$

$$B(t)\mathbf{u}(t) = \tilde{\mathbf{b}}(t) \approx \sum_{i=0}^{m-1} \tilde{\mathbf{b}}_i\phi_i(t) = \tilde{B}\boldsymbol{\phi}(t) \tag{8.77}$$

$$C(t)\mathbf{x}(t-\tau) = \tilde{\mathbf{c}}^\star(t) \approx \sum_{i=0}^{m-1} \tilde{\mathbf{c}}_i^\star(\tau)\phi_i(t) = \tilde{C}^\star(\tau)\boldsymbol{\phi}(t) \tag{8.78}$$

$$K(t)\mathbf{u}(t-\theta) = \tilde{\mathbf{k}}^\star(t) \approx \sum_{i=0}^{m-1} \tilde{\mathbf{k}}_i^\star(\theta)\phi_i(t) = \tilde{K}^\star(\theta)\boldsymbol{\phi}(t) \tag{8.79}$$

$$G(t)\mathbf{x}(t_0 + t_f - t) = \bar{\mathbf{g}}(t) \approx \sum_{i=0}^{m-1} \bar{\mathbf{g}}_i\phi_i(t) = \bar{G}\boldsymbol{\phi}(t) \tag{8.80}$$

$$P(t)\mathbf{u}(t_0 + t_f - t) = \bar{\mathbf{p}}(t) \approx \sum_{i=0}^{m-1} \bar{\mathbf{p}}_i\phi_i(t) = \bar{P}\boldsymbol{\phi}(t) \tag{8.81}$$

where $\boldsymbol{\phi}(t)$ is given in Eq. (7.12) which is either $\mathbf{B}(t)$ or $\mathbf{L}(t)$. Substituting Eqs. (8.6), (8.8) and (8.76)–(8.81) into Eq. (8.73), and using the forward integration operational property in Eq. (2.10)/(2.50) of OFs yield

$$X - V = \left[\tilde{A} + \tilde{B} + \tilde{C}^\star + \tilde{K}^\star + \bar{G} + \bar{P}\right] H$$

$$\Rightarrow \quad \hat{\mathbf{x}} = \hat{\mathbf{v}} + \left(H^T \otimes I_n\right)\left[\hat{\mathbf{a}} + \hat{\mathbf{b}} + \hat{\mathbf{c}}^\star(\tau) + \hat{\mathbf{k}}^\star(\theta) + \hat{\mathbf{g}} + \hat{\mathbf{p}}\right] \tag{8.82}$$

where

$$\hat{\mathbf{x}} = \begin{bmatrix} \mathbf{x}_0 \\ \mathbf{x}_1 \\ \vdots \\ \mathbf{x}_{m-1} \end{bmatrix}; \ \hat{\mathbf{v}} = \begin{bmatrix} \mathbf{v}_0 \\ \mathbf{v}_1 \\ \vdots \\ \mathbf{v}_{m-1} \end{bmatrix}; \ \hat{\mathbf{a}} = \begin{bmatrix} \tilde{\mathbf{a}}_0 \\ \tilde{\mathbf{a}}_1 \\ \vdots \\ \tilde{\mathbf{a}}_{m-1} \end{bmatrix}; \ \hat{\mathbf{b}} = \begin{bmatrix} \tilde{\mathbf{b}}_0 \\ \tilde{\mathbf{b}}_1 \\ \vdots \\ \tilde{\mathbf{b}}_{m-1} \end{bmatrix};$$

$$\hat{\mathbf{c}}^\star(\tau) = \begin{bmatrix} \tilde{\mathbf{c}}_0^\star(\tau) \\ \tilde{\mathbf{c}}_1^\star(\tau) \\ \vdots \\ \tilde{\mathbf{c}}_{m-1}^\star(\tau) \end{bmatrix}; \ \hat{\mathbf{k}}^\star(\theta) = \begin{bmatrix} \tilde{\mathbf{k}}_0^\star(\theta) \\ \tilde{\mathbf{k}}_1^\star(\theta) \\ \vdots \\ \tilde{\mathbf{k}}_{m-1}^\star(\theta) \end{bmatrix}; \ \hat{\mathbf{g}} = \begin{bmatrix} \bar{\mathbf{g}}_0 \\ \bar{\mathbf{g}}_1 \\ \vdots \\ \bar{\mathbf{g}}_{m-1} \end{bmatrix}; \ \hat{\mathbf{p}} = \begin{bmatrix} \bar{\mathbf{p}}_0 \\ \bar{\mathbf{p}}_1 \\ \vdots \\ \bar{\mathbf{p}}_{m-1} \end{bmatrix}$$

$$(8.83)$$

$$\hat{\mathbf{a}} = \hat{A}\hat{\mathbf{x}}, \quad \hat{\mathbf{b}} = \hat{B}\hat{\mathbf{u}}, \quad \hat{\mathbf{c}}^\star(\tau) = \hat{C}\hat{\mathbf{x}}^\star(\tau)$$
$$\hat{\mathbf{k}}^\star(\theta) = \hat{K}\hat{\mathbf{u}}^\star(\theta), \quad \hat{\mathbf{g}} = \hat{G}\bar{I}_n\hat{\mathbf{x}}, \quad \hat{\mathbf{p}} = \hat{P}\bar{I}_r\hat{\mathbf{u}} \qquad (8.84)$$

$$\hat{\mathbf{x}}^\star(\tau) = \begin{bmatrix} \mathbf{x}_0^\star(\tau) \\ \mathbf{x}_1^\star(\tau) \\ \vdots \\ \mathbf{x}_{m-1}^\star(\tau) \end{bmatrix}; \quad \hat{\mathbf{u}} = \begin{bmatrix} \mathbf{u}_0 \\ \mathbf{u}_1 \\ \vdots \\ \mathbf{u}_{m-1} \end{bmatrix}; \quad \hat{\mathbf{u}}^\star(\theta) = \begin{bmatrix} \mathbf{u}_0^\star(\theta) \\ \mathbf{u}_1^\star(\theta) \\ \vdots \\ \mathbf{u}_{m-1}^\star(\theta) \end{bmatrix} \qquad (8.85)$$

and $\hat{A}$, $\hat{B}$, $\hat{C}$, $\hat{K}$, $\hat{G}$, $\hat{P}$ and $\bar{I}$ are defined in the following subsections.

Eq. (8.82) can be rewritten in the form of Eq. (8.19) where M and $\hat{\mathbf{w}}$ are defined in the following subsections.

Now considering the cost function, we express $Q(t)$ and $R(t)$ in terms of OFs as shown in Eqs (8.21) and (8.22). Then the cost function becomes

$$J = \frac{1}{2} \int_{t_0}^{t_f} \left[\mathbf{x}^T(t)Q(t)\mathbf{x}(t) + \mathbf{u}^T(t)R(t)\mathbf{u}(t) \right] dt$$
$$\simeq \frac{1}{2} \left(\hat{\mathbf{x}}^T \hat{Q}\hat{\mathbf{x}} + \hat{\mathbf{u}}^T \hat{R}\hat{\mathbf{u}} \right) \qquad (8.86)$$

where $\hat{Q}$ and $\hat{R}$ are defined in the following subsections.

Substituting Eq. (8.19) into Eq. (8.86) and setting the optimization condition in Eq. (7.41) yield the optimal control law [91]

$$\hat{\mathbf{u}} = -\left[M^T \hat{Q} M + \hat{R}\right]^{-1} M^T \hat{Q} \hat{\mathbf{w}} \qquad (8.87)$$

8.3.1 Using BPFs

In Eq. (8.82) $\hat{\mathbf{v}}$ is as given in Eq. (8.25). In Eq. (8.84), we have [91]

$$\hat{A} = \text{diag}\left[\begin{array}{cccc} A_0, & A_1, & \ldots, & A_{m-1} \end{array}\right] \qquad (8.88)$$

$$\hat{B} = \text{diag}\left[\begin{array}{cccc} B_0, & B_1, & \ldots, & B_{m-1} \end{array}\right] \qquad (8.89)$$

$$\hat{C} = \text{diag}\left[\begin{array}{cccc} C_0, & C_1, & \ldots, & C_{m-1} \end{array}\right] \qquad (8.90)$$

$$\hat{K} = \text{diag}\left[\begin{array}{cccc} K_0, & K_1, & \ldots, & K_{m-1} \end{array}\right] \qquad (8.91)$$

$$\hat{G} = \text{diag}\left[\begin{array}{cccc} G_0, & G_1, & \ldots, & G_{m-1} \end{array}\right] \qquad (8.92)$$

$$\hat{P} = \text{diag}\left[\begin{array}{cccc} P_0, & P_1, & \ldots, & P_{m-1} \end{array}\right] \qquad (8.93)$$

$$\bar{I}_n = \begin{bmatrix} O & O & \cdots & O & I_n \\ O & O & \cdots & I_n & O \\ \vdots & \vdots & & \vdots & \vdots \\ O & I_n & \cdots & O & O \\ I_n & O & \cdots & O & O \end{bmatrix} ;$$

$$\bar{I}_r = \begin{bmatrix} O & O & \cdots & O & I_r \\ O & O & \cdots & I_r & O \\ \vdots & \vdots & & \vdots & \vdots \\ O & I_r & \cdots & O & O \\ I_r & O & \cdots & O & O \end{bmatrix} \qquad (8.94)$$

which are $mn \times mn$ and $mr \times mr$ matrices respectively,

$$\hat{\mathbf{x}}^\star(\tau) = E\left(n, \mu\right) \hat{\boldsymbol{\zeta}}(\tau) + D\left(n, \mu\right) \hat{\mathbf{x}} \qquad (8.95)$$

$$\hat{\mathbf{u}}^\star(\theta) = E\left(r, \delta\right) \hat{\boldsymbol{\nu}}(\theta) + D\left(r, \delta\right) \hat{\mathbf{u}} \qquad (8.96)$$

where

$$\tau = \mu T \quad \text{and} \quad \theta = \delta T \tag{8.97}$$

with μ and δ being the number of BPFs on $t_0 \leq t \leq t_0 + \tau$ and $t_0 \leq t \leq t_0 + \theta$, respectively,

$$\hat{\zeta}(\tau) = \begin{bmatrix} \zeta_0(\tau) \\ \zeta_1(\tau) \\ \vdots \\ \zeta_{\mu-1}(\tau) \end{bmatrix} ; \quad \hat{\nu}(\theta) = \begin{bmatrix} \nu_0(\theta) \\ \nu_1(\theta) \\ \vdots \\ \nu_{\delta-1}(\theta) \end{bmatrix} \tag{8.98}$$

$$\zeta_i(\tau) = \frac{1}{T} \int_{t_0+iT}^{t_0+(i+1)T} \zeta(t-\tau)dt \quad \text{for } i = 0,1,2,\ldots,\mu-1 \tag{8.99}$$

$$\text{and } \nu_i(\theta) = \frac{1}{T} \int_{t_0+iT}^{t_0+(i+1)T} \nu(t-\theta)dt \quad \text{for } i = 0,1,2,\ldots,\delta-1 \tag{8.100}$$

In Eq. (8.19)

$$M = N^{-1} \left(H^T \otimes I_n \right) \left[\hat{B} + \hat{K}D(r,\ \delta) + \hat{P}\bar{I}_r \right] \tag{8.101}$$

$$\hat{\mathbf{w}} = N^{-1} \left\{ \hat{\mathbf{v}} + \left(H^T \otimes I_n \right) \left[\hat{C}E(n,\ \mu)\hat{\zeta}(\tau) + \hat{K}E(r,\ \delta)\hat{\nu}(\theta) \right] \right\} \tag{8.102}$$

where

$$N = I_{mn} - \left(H^T \otimes I_n \right) \left[\hat{A} + \hat{C}D(n,\ \mu) + \hat{G}\bar{I}_n \right] \tag{8.103}$$

In Eq. (8.86) $\hat{Q}$ and $\hat{R}$ are as given in Eqs. (8.38) and (8.39), respectively.

8.3.2 Using SLPs

In Eq. (8.82) $\hat{\mathbf{v}}$ is as given in Eq. (8.40). In Eq. (8.84), we have [91]

$$
\hat{A} = \frac{1}{(t_f - t_0)}
\begin{bmatrix}
\sum\limits_{j=0}^{m-1} \pi_{0j\,0} A_j & \cdots & \sum\limits_{j=0}^{m-1} \pi_{0j,m-1} A_j \\
3 \sum\limits_{j=0}^{m-1} \pi_{1j\,0} A_j & \cdots & 3 \sum\limits_{j=0}^{m-1} \pi_{1j,m-1} A_j \\
\vdots & & \vdots \\
(2m-1) \sum\limits_{j=0}^{m-1} \pi_{m-1,j\,0} A_j & \cdots & (2m-1) \sum\limits_{j=0}^{m-1} \pi_{m-1,j,m-1} A_j
\end{bmatrix}
\tag{8.104}
$$

$\hat{B}$, $\hat{C}$, $\hat{K}$, $\hat{G}$ and $\hat{P}$ are similar to $\hat{A}$ in Eq. (8.104),

$$
\bar{I}_n =
\begin{bmatrix}
I_n & O & O & \cdots & O \\
O & -I_n & O & \cdots & O \\
O & O & I_n & \cdots & O \\
\vdots & \vdots & \vdots & & \vdots \\
O & O & O & \cdots & (-1)^{m-1} I_n
\end{bmatrix} ;
$$

$$
\bar{I}_r =
\begin{bmatrix}
I_r & O & O & \cdots & O \\
O & -I_r & O & \cdots & O \\
O & O & I_r & \cdots & O \\
\vdots & \vdots & \vdots & & \vdots \\
O & O & O & \cdots & (-1)^{m-1} I_r
\end{bmatrix}
\tag{8.105}
$$

$$
\hat{\mathbf{x}}^{\star}(\tau) = \hat{\boldsymbol{\zeta}}(\tau) + (D(\tau) \otimes I_n)\,\hat{\mathbf{x}}
\tag{8.106}
$$

$$
\hat{\mathbf{u}}^{\star}(\theta) = \hat{\boldsymbol{\nu}}(\theta) + (D(\theta) \otimes I_r)\,\hat{\mathbf{u}}
\tag{8.107}
$$

where

$$
\hat{\boldsymbol{\zeta}}(\tau) =
\begin{bmatrix}
\boldsymbol{\zeta}_0(\tau) \\
\boldsymbol{\zeta}_1(\tau) \\
\vdots \\
\boldsymbol{\zeta}_{m-1}(\tau)
\end{bmatrix}
; \quad
\hat{\boldsymbol{\nu}}(\theta) =
\begin{bmatrix}
\boldsymbol{\nu}_0(\theta) \\
\boldsymbol{\nu}_1(\theta) \\
\vdots \\
\boldsymbol{\nu}_{m-1}(\theta)
\end{bmatrix}
\tag{8.108}
$$

$$\zeta_i(\tau) = \frac{(2i+1)}{(t_f - t_0)} \int_{t_0}^{t_0+\tau} \zeta(t-\tau)L_i(t)dt \qquad (8.109)$$

$$\nu_i(\theta) = \frac{(2i+1)}{(t_f - t_0)} \int_{t_0}^{t_0+\theta} \nu(t-\theta)L_i(t)dt \qquad (8.110)$$

for $i = 0, 1, 2, \ldots, m-1$. In Eq. (8.19)

$$M = N^{-1} \left(H^T \otimes I_n \right) \left[\hat{B} + \hat{K} \left(D(\theta) \otimes I_r \right) + \hat{P}\bar{I}_r \right] \quad (8.111)$$

$$\hat{\mathbf{w}} = N^{-1} \left\{ \hat{\mathbf{v}} + \left(H^T \otimes I_n \right) \left[\hat{C}\hat{\zeta}(\tau) + \hat{K}\hat{\nu}(\theta) \right] \right\} \qquad (8.112)$$

where

$$N = I_{mn} - \left(H^T \otimes I_n \right) \left[\hat{A} + \hat{C} \left(D(\tau) \otimes I_n \right) + \hat{G}\bar{I}_n \right] (8.113)$$

In Eq. (8.86) $\hat{Q}$ and $\hat{R}$ are as given in Eqs. (8.54) and (8.55), respectively.

8.3.3 Illustrative example

Consider the system [73] described by

$$\begin{bmatrix} \dot{x}_1(t) \\ \dot{x}_2(t) \\ \dot{x}_3(t) \end{bmatrix} = \begin{bmatrix} 1 & t+1 \\ t+1 & 1 \\ 1 & 1 \end{bmatrix} \begin{bmatrix} u_1(t) \\ u_2(t) \end{bmatrix}$$

$$+ \begin{bmatrix} t & 1 & t^2+1 \\ 1 & t & 0 \\ t^2+1 & 0 & t \end{bmatrix} \begin{bmatrix} x_1\left(t-\frac{1}{3}\right) \\ x_2\left(t-\frac{1}{3}\right) \\ x_3\left(t-\frac{1}{3}\right) \end{bmatrix}$$

$$+ \begin{bmatrix} 1 & t+1 \\ t & 1 \\ 1 & t^2+1 \end{bmatrix} \begin{bmatrix} u_1\left(t-\frac{2}{3}\right) \\ u_2\left(t-\frac{2}{3}\right) \end{bmatrix}$$

$$+ \begin{bmatrix} t^2 & 1 & 1 \\ 1 & 1 & 0 \\ t & 0 & 1 \end{bmatrix} \begin{bmatrix} x_1(1-t) \\ x_2(1-t) \\ x_3(1-t) \end{bmatrix}$$

$$\begin{bmatrix} x_1(t) \\ x_2(t) \\ x_3(t) \end{bmatrix} = \begin{bmatrix} 1, & 1, & 1 \end{bmatrix}^T \quad \text{for} \quad -\frac{1}{3} \leq t \leq 0$$

$$\begin{bmatrix} u_1(t) \\ u_2(t) \end{bmatrix} = \begin{bmatrix} 1, & 1 \end{bmatrix}^T \quad \text{for} \quad -\frac{2}{3} \leq t \leq 0$$

with the cost function

$$
J = \frac{1}{2} \int_0^1 \left\{ \left[x_1(t), \ x_2(t), \ x_3(t) \right] \begin{bmatrix} 1 & t & 0 \\ t & t^2 & 0 \\ 0 & 0 & t^2 \end{bmatrix} \begin{bmatrix} x_1(t) \\ x_2(t) \\ x_3(t) \end{bmatrix} \right.
$$

$$
\left. + \left[u_1(t) \ u_2(t) \right] \begin{bmatrix} u_1(t) \\ u_2(t) \end{bmatrix} \right\} dt
$$

Since $t_f = 1$, $\tau = \frac{1}{3}$ and $\theta = \frac{2}{3}$, $m = 18$ for the BPF approach and $m = 12$ for the SLP approach are chosen and computed $\mathbf{x}(t)$, $\mathbf{u}(t)$ and J. $u_1(t)$ and $u_2(t)$ are shown in Figs. 8.8 and 8.9, and $x_1(t)$, $x_2(t)$ and $x_3(t)$ are shown in Fig. 8.10. The J value is shown in Table 8.6. For comparison, the results obtained by Wang [73] are also reproduced (Case : # of BPFs = 3 and # of SLPs = 6). It is interesting to note from the figures and table that the results obtained via BPFs and SLPs are consistent with each other while it is not so with hybrid functions.

Table 8.6: Cost function

Method	m	J
BPF	18	3.1092085758
SLP	12	3.1070469742
Hybrid functions [73]	# BPFs = 3, # SLPs = 6	3.3693347738

8.4 Conclusion

Two unified approaches for computing optimal control law have been presented; one [87, 90, 94] for linear time-invariant/ time-varying systems with/without time delays in control and state, and the other [91] for linear time-varying systems with time-delay and reverse time terms in state and control. Our methods are based on using two classes of OFs, namely BPFs and SLPs. The nature of OFs used is reflected in the final solution, i.e. the solution is

156

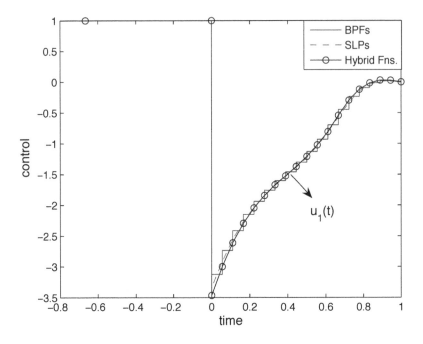

Figure 8.8: BPF, SLP and hybrid functions solutions of $u_1(t)$.

always piecewise constant if BPFs (piecewise constant functions) are used while it is smooth with SLPs (polynomial functions). The illustrative examples demonstrated the validity and effectiveness of the approaches.

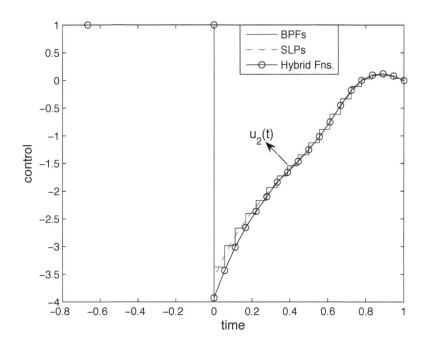

Figure 8.9: BPF, SLP and hybrid functions solutions of $u_2(t)$.

158

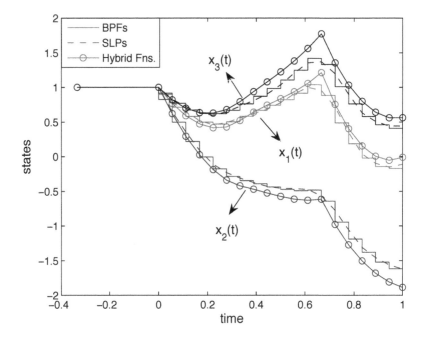

Figure 8.10: BPF, SLP and hybrid functions solutions of state variables.

Chapter 9

Optimal Control of Nonlinear Systems

Using the nonlinear operational matrix derived in Chapter 2 and other relevant operational properties of LPs and BPFs, an approximate solution for a nonlinear optimal control problem with quadratic performance index is given. Three illustrative examples are included to demonstrate the validity of the approach.

9.1 Introduction

Looking at the historical developments on solving optimal control problem of nonlinear systems via the OF approach, we find that not much work has been reported. Lee and Chang [40] appear to be the first to study the optimal control problem of nonlinear systems using GOPs. For this, they introduced a nonlinear operational matrix of GOPs. Though their work is fundamental and significant, it is not attractive computationally, as it involves the Kronecker product [12] operation.

Chebyshev polynomials were used [49] for solving the nonlinear optimal control problem. In [54] a general framework for the

159

nonlinear optimal control problem was developed by employing BPFs. Using quasilinearisation and Chebyshev polynomials, a numerical method was presented [65] to solve the nonlinear optimal control problem. Very recently, for solving optimal control problems an efficient algorithm [93] has been developed by employing a fast Chebyshev transform and a fast Legendre transform. Both control and state functions have been approximated in terms of Chebyshev expansions.

To have more computational simplicity, in this chapter, we employ BPFs and LPs for computing the optimal control law of nonlinear systems. Using the nonlinear operational matrix derived in Chapter 2 and other relevant properties of LPs and BPFs, a computationally elegant approach [95] is presented for finding the optimal control law of nonlinear systems.

The chapter is organized as follows: The optimal control problem is considered in Section 9.2. Section 9.3 includes three illustrative examples. The last section concludes the chapter.

9.2 Computation of the Optimal Control Law

The aim is to compute the optimal control law $u(t)$ that minimises the cost function

$$J = \frac{1}{2} \int_{t_0}^{t_f} \left[q_1 x_1^2(t) + q_2 x_2^2(t) + ru^2(t) \right] dt \qquad (9.1)$$

subject to the constraint

$$\dot{x}_i(t) = f_i(x_1(t), x_2(t)) + g_i(x_1(t), x_2(t)) u(t) \qquad (9.2)$$

with the initial conditions $x_i(t_0) = \tilde{x}_i$ where $i = 1$ and 2, $q_i \geq 0$, $r > 0$, and t_0 and t_f are the initial and final times.

Integrating Eq. (9.2) with respect to t, we get

$$x_i(t) - x_i(t_0) = \int_{t_0}^{t} [f_i(x_1(\tau), x_2(\tau)) + g_i(x_1(\tau), x_2(\tau)) u(\tau)] d\tau$$

(9.3)

Expressing all the terms in Eq. (9.3) in terms of OFs,

$$x_i(t) \approx \sum_{j=0}^{m-1} x_{ij} B_j(t) = \mathbf{x}_i^T \mathbf{B}(t)$$

(9.4)

$$x_i(t_0) \approx x_i(t_0) \sum_{j=0}^{m-1} B_j(t) = \tilde{\mathbf{x}}_i^T \mathbf{B}(t)$$

(9.5)

Representation of $f(x(t), y(t))$ in OFs was shown in Eq. (2.104). f_1, f_2, g_1 and g_2 can also be represented in the similar manner. Now consider

$$g(x(t), y(t)) u(t) \approx \sum_{i=0}^{n-1} \sum_{j=0}^{n-1} \sum_{k=0}^{m-1} n_{ik}(x) g_{ij} n_{jk}(y) B_k(t) \sum_{l=0}^{m-1} u_l B_l(t)$$

$$\approx \sum_{i=0}^{n-1} \sum_{j=0}^{n-1} \sum_{k=0}^{m-1} n_{ik}(x) g_{ij} n_{jk}(y) u_k B_k(t) \quad (9.6)$$

Therefore, $g_1 u$ and $g_2 u$ can also be represented in OFs. Substituting all the terms into Eq. (9.3), making use of the integration operational property of BPFs in Eq. (2.10), and simplifying, lead to

$$x_{i0} = \tilde{x}_i + 0.5T \sum_{j=0}^{n-1} \sum_{k=0}^{n-1} n_{j0}(x) (f_{ijk} + g_{ijk} u_0) n_{k0}(y) \quad (9.7)$$

$$x_{il} - x_{i,l-1} = 0.5T \sum_{j=0}^{n-1} \sum_{k=0}^{n-1} [n_{j,l-1}(x) (f_{ijk} + g_{ijk} u_{l-1}) n_{k,l-1}(y)$$

$$+ n_{jl}(x) (f_{ijk} + g_{ijk} u_l) n_{kl}(y)] \quad (9.8)$$

for $l = 1, 2, \ldots, m - 1$.

Now expressing $x_i^2(t)$ and $u^2(t)$ in terms of LPs and BPFs, we have

$$x_i^2(t) = \sum_{j=0}^{2} c_j P_j(x_i(t)) \approx \sum_{j=0}^{2} c_j \sum_{k=0}^{m-1} n_{jk}(x_i) B_k(t) \quad (9.9)$$

$$u^2(t) = \sum_{j=0}^{2} c_j P_j(u(t)) \approx \sum_{j=0}^{2} c_j \sum_{k=0}^{m-1} n_{jk}(u) B_k(t) \quad (9.10)$$

where c_j is the jth Legendre spectrum element of $x_1^2(t)$, $x_2^2(t)$ or $u^2(t)$. Substituting Eqs. (9.9) and (9.10) into Eq. (9.1) and using integration operational property of BPFs, we obtain

$$J = \frac{T}{2} \sum_{j=0}^{2} c_j \sum_{k=0}^{m-1} [q_1 n_{jk}(x_1) + q_2 n_{jk}(x_2) + r n_{jk}(u)] \quad (9.11)$$

Thus, the optimal control problem becomes minimization of Eq. (9.11) subject to the constraint in Eqs. (9.7) and (9.8). This non-linear optimization problem can be solved for the spectra of $x_i(t)$ and $u(t)$ using Newton's iterative method.

9.3 Illustrative Examples

Example 9.1

Find the optimal control law $u(t)$ which minimizes

$$J = \frac{1}{2} \int_0^1 [x^2(t) + u^2(t)] \, dt \quad (9.12)$$

such that the bilinear system dynamics

$$\dot{x}(t) = -2x(t) + x(t) u(t) + 3u(t), \quad x(0) = 5$$

are satisfied [54].

This problem is solved using our approach with $m = 10$ and $n = 3$. Figures 9.1 and 9.2 show the computed $u(t)$ and $x(t)$.

For comparison, the exact solution [14] is also shown in the same figures. It is clear from the figures that the solution obtained via our approach follows the exact solution. The J value is shown in Table 9.1.

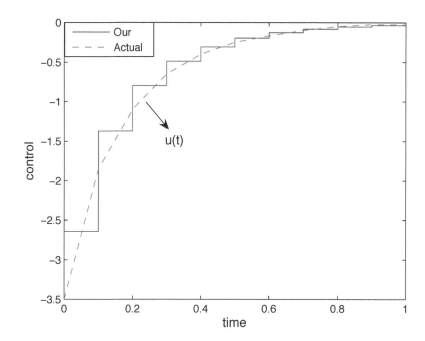

Figure 9.1: Control variable in Example 1.

Example 9.2

Find optimal control law $u(t)$ which minimizes Eq. (9.12) while satisfying the dynamics of the nonlinear system [54]

$$\dot{x}(t) = -x^2(t) + u(t), \quad x(0) = 10$$

This problem is solved with $m = 10$ and $n = 3$, and the results obtained are shown in Figures 9.3 and 9.4, which also show the exact solution [14]. The results are satisfactory. The J value is shown in Table 9.1.

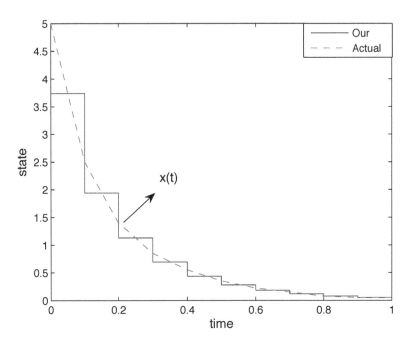

Figure 9.2: State variable in Example 1.

Example 9.3

Consider the second-order nonlinear system [10, 40]

$$\dot{x}_1(t) = x_2(t) + 0.01x_2^3(t), \quad x_1(0) = 2$$
$$\dot{x}_2(t) = -4x_1(t) - 5x_2(t) + 4u(t), \quad x_2(0) = 0$$

It is desired to find the optimal control law of the form $u(t) = -kx_1(t)$ where the feedback gain k is to be chosen such that

$$J = \frac{1}{2} \int_0^1 \left[x_1^2(t) + u^2(t) \right] dt$$

is minimum.

This problem is solved with $m = 6$ and $n = 4$, and the results are shown in Figures 9.5 and 9.6. The feedback gain is found

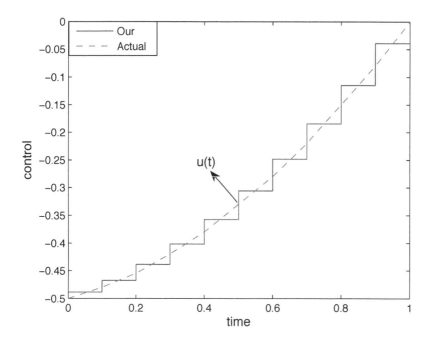

Figure 9.3: Control variable in Example 2.

to be $k = 0.2334$. The results obtained via the quasilinearization technique [10] and Runge-Kutta method are also shown for
comparison. It can be seen that our approach yields satisfactory
results. The value of J is shown in Table 9.1.

Table 9.1: Cost function

Example	m	n	J
1	10	3	1.4851
2	10	3	4.4951
3	6	4	1.1439
3	Sage & White	method [10]	1.7527

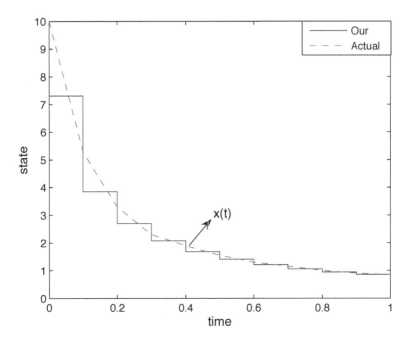

Figure 9.4: State variable in Example 2.

9.4 Conclusion

The derived nonlinear operational matrix is computationally simpler than the one obtained by Lee and Chang [40] as it is completely recursive in nature and free from the Kronecker product [12] operation. Moreover, the approach to find optimal control law of nonlinear systems is also computationally simpler than that of [40] due to its nondependence on the Kronecker product operation. The simplicity in computation is achieved by means of the disjoint property of BPFs. Thus, our method is a very convenient and tractable method for solving nonlinear optimal control problems numerically.

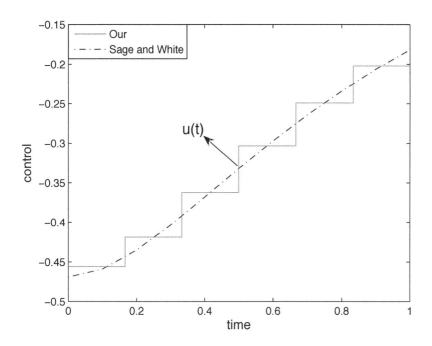

Figure 9.5: Control variable in Example 3.

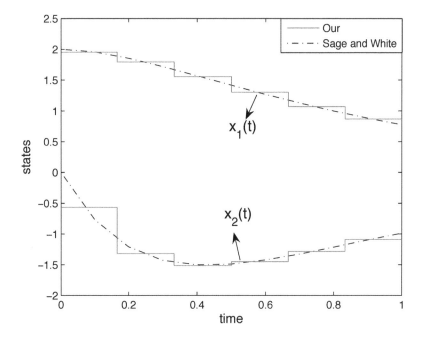

Figure 9.6: State variables in Example 3.

Chapter 10

Hierarchical Control of Linear Systems

This chapter presents a BPF approach to hierarchical control of linear time-invariant/time-varying systems with quadratic cost functions. Owing to the elegant operational properties of BPFs, the computational formulations for the interaction prediction method are shown to be purely algebraic. The resulting solutions are piecewise constant with minimal mean-square error. The suggested approach is computationally simple and attractive. Three illustrative examples are included to show the applicability of the approach.

10.1   Introduction

One may use the well known control paradigms like hierarchical control and decentralized control for controlling large scale systems if their mathematical models are available. Success of the control solution depends on the computational technique employed. A number of reliable techniques are available today.

Hierarchical control problems have been studied via BPFs [45].

We discuss this BPF approach in this chapter, which is organized as follows: The next section briefly touches upon hierarchical control of linear time-invariant systems. Section 3 discusses solution of the hierarchical control problem via a BPF approach. Extension to linear time-varying systems is presented in Section 4 while Section 5 contains computational algorithm. Three illustrative examples are included in Section 6. The last section concludes the chapter.

10.2 Hierarchical Control of LTI Systems with Quadratic Cost Functions

Fig. 10.1 shows a hierarchical control structure. The large scale system contains systems S_1 and S_2 which in turn contain subsystems s_{11}, s_{12}, s_{13} and s_{21}, s_{22}, s_{23}, s_{24} respectively. Each subsystem s_{ij} has its controller (may be a linear-quadratic-regulator) to take care of its own behaviour. Subcoordinators C_1 and C_2 take care of the interactions among the subsystems s_{11}, s_{12} and s_{13}, and among the subsystems s_{21}, s_{22}, s_{23} and s_{24} through an iterative process of modeling. Coordinator C similarly takes care of the coordination between systems S_1 and S_2. Thus a convergent optimal solution can be obtained towards the control of a large scale system.

Let us assume that the large scale system comprises N systems which are interconnected. For any system S_i, $\mathbf{x}_i$ is the n_i dimensional state vector, $\mathbf{u}_i$ is the m_i dimensional control vector, and $\mathbf{z}_i$ is the q_i dimensional vector of interconnections from the other systems S_j, $j = 1, 2, \ldots, i - 1, i + 1, \ldots, N$. We assume that the

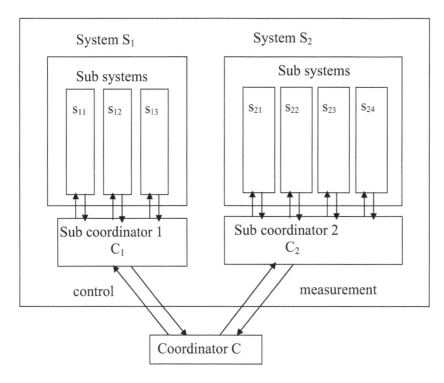

Figure 10.1: Hierarchical control strategy for a large scale system.

systems themselves can be described by

$$\dot{\mathbf{x}}_i(t) = A_i\mathbf{x}_i(t) + B_i\mathbf{u}_i(t) + \mathbf{z}_i(t); \quad \mathbf{x}_i(t_0) = \mathbf{x}_{i0} \quad (10.1)$$

We assume also that the vector of inputs $\mathbf{z}_i$ is a linear combination of the states of the $(N-1)$ systems, i.e.

$$\mathbf{z}_i = \sum_{\substack{j=1 \\ j \neq i}}^{N} L_{ij}\mathbf{x}_j \quad (10.2)$$

The problem of interest is to choose the controls $\mathbf{u}_1, \ldots, \mathbf{u}_N$ in order to minimize the quadratic cost function

$$J = \frac{1}{2}\sum_{i=1}^{N}\int_{t_0}^{t_f}\left[\mathbf{x}_i^T(t)Q_i\mathbf{x}_i(t) + \mathbf{u}_i^T(t)R_i\mathbf{u}_i(t)\right]dt \quad (10.3)$$

subject to the constraints in Eqs. (10.1) and (10.2) where Q_i is a positive semi-definite matrix, R_i is a positive definite matrix, and t_0 and t_f are the initial and final times.

10.2.1 Partial feedback control

The Hamiltonian for the ith independent problem can be written [13, 21] as

$$
\begin{aligned}
H_i &= \frac{1}{2}\left[\mathbf{x}_i^T Q_i \mathbf{x}_i + \mathbf{u}_i^T R_i \mathbf{u}_i\right] + \boldsymbol{\lambda}_i^T \mathbf{z}_i \\
&\quad - \sum_{\substack{j=1 \\ j \neq i}}^{N} \boldsymbol{\lambda}_j^T L_{ji} \mathbf{x}_i + \mathbf{p}_i^T \left[A_i \mathbf{x}_i + B_i \mathbf{u}_i + \mathbf{z}_i\right]
\end{aligned} \tag{10.4}
$$

where $\boldsymbol{\lambda}_i$ is the $\mathbf{q}_i$ dimensional vector of Lagrange multipliers, and $\mathbf{p}_i$ is the n_i dimensional adjoint vector. Then from the necessary conditions

$$
\frac{\partial H_i}{\partial \mathbf{x}_i} = -\dot{\mathbf{p}}_i \quad \text{and} \quad \frac{\partial H_i}{\partial \mathbf{u}_i} = 0
$$

we have

$$
\dot{\mathbf{p}}_i = -Q_i \mathbf{x}_i - A_i^T \mathbf{p}_i + \sum_{\substack{j=1 \\ j \neq i}}^{N} L_{ji}^T \boldsymbol{\lambda}_j \; ; \quad \mathbf{p}_i(t_f) = \mathbf{0} \tag{10.5}
$$

$$
\mathbf{u}_i = -R_i^{-1} B_i^T \mathbf{p}_i \tag{10.6}
$$

Let

$$
\mathbf{p}_i = K_i \mathbf{x}_i + \mathbf{s}_i \tag{10.7}
$$

where K_i is the $n_i \times n_i$ symmetric positive definite Riccati matrix, and $\mathbf{s}_i$ is the n_i dimensional open-loop compensation vector. Then

$$
\dot{\mathbf{p}}_i = K_i \dot{\mathbf{x}}_i + \dot{K}_i \mathbf{x}_i + \dot{\mathbf{s}}_i \tag{10.8}
$$

Substituting Eq. (10.6) into Eq. (10.1) and making use of Eq. (10.7), we obtain

$$\dot{\mathbf{x}}_i = A_i \mathbf{x}_i - B_i R_i^{-1} B_i^T \left(K_i \mathbf{x}_i + \mathbf{s}_i \right) + \mathbf{z}_i ; \quad \mathbf{x}_i(t_0) = \mathbf{x}_{i0} \quad (10.9)$$

Substituting Eqs. (10.5) and (10.9) into Eq. (10.8), we have

$$-Q_i \mathbf{x}_i - A_i^T \left(K_i \mathbf{x}_i + \mathbf{s}_i \right) + \sum_{\substack{j=1 \\ j \neq i}}^{N} L_{ji}^T \boldsymbol{\lambda}_j = K_i [A_i \mathbf{x}_i$$

$$- B_i R_i^{-1} B_i^T \left(K_i \mathbf{x}_i + \mathbf{s}_i \right) + \mathbf{z}_i] + \dot{K}_i \mathbf{x}_i + \dot{\mathbf{s}}_i \quad (10.10)$$

which is valid for arbitrary $\mathbf{x}_i$, leading to

$$\dot{K}_i + K_i A_i + A_i^T K_i - K_i B_i R_i^{-1} B_i^T K_i + Q_i = \bigcirc ; \quad K_i(t_f) = \bigcirc \quad (10.11)$$

and

$$\dot{\mathbf{s}}_i = -K_i \mathbf{z}_i - \left(A_i - B_i R_i^{-1} B_i^T K_i \right)^T \mathbf{s}_i + \sum_{\substack{j=1 \\ j \neq i}}^{N} L_{ji}^T \boldsymbol{\lambda}_j ; \quad \mathbf{s}_i(t_f) = \mathbf{0} \quad (10.12)$$

The local control law is given by

$$\mathbf{u}_i = -R_i^{-1} B_i^T \left(K_i \mathbf{x}_i + \mathbf{s}_i \right) \quad (10.13)$$

10.2.2 Interaction prediction approach

Let us write the Lagrangian [13, 21] as

$$L(\mathbf{x}_i, \mathbf{u}_i, \boldsymbol{\lambda}_i, \mathbf{z}_i, \mathbf{p}_i) = \sum_{i=1}^{N} L_i$$

$$= \sum_{i=1}^{N} \int_{t_0}^{t_f} [0.5 \left(\mathbf{x}_i^T Q_i \mathbf{x}_i + \mathbf{u}_i^T R_i \mathbf{u}_i \right) + \boldsymbol{\lambda}_i^T \mathbf{z}_i$$

$$- \sum_{\substack{j=1 \\ j \neq i}}^{N} \boldsymbol{\lambda}_j^T L_{ji} \mathbf{x}_i + \mathbf{p}_i^T \left(-\dot{\mathbf{x}}_i + A_i \mathbf{x}_i + B_i \mathbf{u}_i + \mathbf{z}_i \right)] dt \quad (10.14)$$

Setting the necessary conditions

$$\frac{\partial L}{\partial \mathbf{z}_i} = \mathbf{0} \quad \text{and} \quad \frac{\partial L}{\partial \boldsymbol{\lambda}_i} = \mathbf{0}$$

gives

$$\boldsymbol{\lambda}_i = -\mathbf{p}_i \tag{10.15}$$

and

$$\mathbf{z}_i = \sum_{\substack{j=1 \\ j \neq i}}^{N} L_{ij} \mathbf{x}_j \tag{10.16}$$

thus making the coordination rule at $(k+1)^{th}$ iteration from the kth iteration

$$\begin{bmatrix} \boldsymbol{\lambda}_i \\ \mathbf{z}_i \end{bmatrix}^{k+1} = \begin{bmatrix} -\mathbf{p}_i \\ \sum_{\substack{j=1 \\ j \neq i}}^{N} L_{ij} \mathbf{x}_j \end{bmatrix}^{k} \tag{10.17}$$

10.3   Solution of Hierarchical Control Problem via BPFs

The K_i in Eq. (10.11) is independent of the initial state $\mathbf{x}_i(0)$. Thus N number of matrix Riccati equations each involving $n_i \times \frac{(n_i+1)}{2}$ nonlinear differential equations are to be solved independently from the given final condition $K_i(t_f) = \bigcirc$. These give a partial feedback control.

Finding partial feedback control via Riccati matrix K_i is not easy computationally as it involves solution of nonlinear differential equations, and the method of obtaining numerical solution of nonlinear differential equations is iterative in nature.

Here we present a method, called the state transition matrix method, which is purely recursive in nature.

Let

$$F_i = \begin{bmatrix} A_i & -B_i R_i^{-1} B_i^T \\ -Q_i & -A_i^T \end{bmatrix}, \quad \text{and} \quad g_i = \sum_{\substack{j=1 \\ j \neq i}}^{N} L_{ji}^T \lambda_j \qquad (10.18)$$

Then Eqs. (10.1), (10.5) and (10.6) can be written as

$$\begin{bmatrix} \dot{x}_i \\ \dot{p}_i \end{bmatrix} = F_i \begin{bmatrix} x_i \\ p_i \end{bmatrix} + \begin{bmatrix} z_i \\ g_i \end{bmatrix} \qquad (10.19)$$

10.3.1   State transition matrix

Let $\Phi_i(t_f, t)$ be the state transition matrix of the system in Eq. (10.19) where

$$\Phi_i(t_f, t) = \begin{bmatrix} \Phi_{i11}(t_f, t) & \Phi_{i12}(t_f, t) \\ \Phi_{i21}(t_f, t) & \Phi_{i22}(t_f, t) \end{bmatrix}, \quad \Phi_i(t_f, t_f) = I_{2n_i} \qquad (10.20)$$

It is well known that $\Phi_i(t_f, t)$ satisfies the matrix differential equation

$$\dot{\Phi}_i(t_f, t) = -\Phi_i(t_f, t) F_i \qquad (10.21)$$

Integrating backward with respect to time t gives

$$\Phi_i(t_f, t) - I_{2n_i} = -\int_{t_f}^{t} \Phi_i(t_f, \tau) \, d\tau \, F_i \qquad (10.22)$$

Let the BPF representation of $\Phi_i(t_f, t)$ be

$$\Phi_i(t_f, t) \simeq \sum_{j=0}^{\mu-1} \Phi_{ij} B_j(t) = \tilde{\Phi}_i \left(B(t) \otimes I_{2n_i} \right) \qquad (10.23)$$

where $\otimes$ is the Kronecker product [12] and

$$\tilde{\Phi}_i = \begin{bmatrix} \Phi_{i0}, & \Phi_{i1}, & \dots, & \Phi_{i, \mu-1} \end{bmatrix} \qquad (10.24)$$

Similarly,

$$I_{2n_i} = \tilde{I}_{2n_i} \left(\boldsymbol{B}(t) \otimes I_{2n_i} \right) \tag{10.25}$$

where

$$\tilde{I}_{2n_i} = \begin{bmatrix} I_{2n_i}, & I_{2n_i}, & \dots, & I_{2n_i} \end{bmatrix} \tag{10.26}$$

Substituting Eqs. (10.23) and (10.25) into Eq. (10.22), and making use of Eq. (2.12) lead to

$$\left(\tilde{\Phi}_i - \tilde{I}_{2n_i} \right) \left(\boldsymbol{B}(t) \otimes I_{2n_i} \right) = -\hat{\Phi}_i \left(H_b \boldsymbol{B}(t) \otimes I_{2n_i} \right)$$
$$= -\hat{\Phi}_i \left(H_b \otimes I_{2n_i} \right) \left(\boldsymbol{B}(t) \otimes I_{2n_i} \right)$$

so that

$$\tilde{\Phi}_i - \tilde{I}_{2n_i} = -\hat{\Phi}_i \left(H_b \otimes I_{2n_i} \right) \tag{10.27}$$

where

$$\hat{\Phi}_i = \begin{bmatrix} \Phi_{i0}F_i, & \Phi_{i1}F_i, & \dots, & \Phi_{i,\mu-1}F_i \end{bmatrix} \tag{10.28}$$

Upon substituting H_b in Eq. (2.13) into Eq. (10.27) and simplifying, we obtain the following recursive relations to evaluate $\Phi_i(t_f, t)$:

$$\Phi_{i,\mu-1} = \left(I_{2n_i} - 0.5TF_i \right)^{-1} \tag{10.29}$$

and

$$\Phi_{i,j} = \Phi_{i,j+1} \left(I_{2n_i} + 0.5TF_i \right) \left(I_{2n_i} - 0.5TF_i \right)^{-1} \tag{10.30}$$
$$\text{for} \quad j = \mu - 2, \mu - 3, \dots, 1, 0.$$

From Eq. (10.19) it is possible to write

$$\begin{bmatrix} \mathbf{x}_i(t_f) \\ \mathbf{p}_i(t_f) \end{bmatrix} = \begin{bmatrix} \mathbf{x}_i(t_f) \\ \mathbf{0} \end{bmatrix} = \Phi_i(t_f, t) \begin{bmatrix} \mathbf{x}_i(t) \\ \mathbf{p}_i(t) \end{bmatrix}$$
$$+ \int_t^{t_f} \Phi_i(t_f, \tau) \begin{bmatrix} \mathbf{z}_i(\tau) \\ \mathbf{g}_i(\tau) \end{bmatrix} d\tau \tag{10.31}$$

from which we have

$$\Phi_{i21}(t_f,\, t)\mathbf{x}_i(t) + \Phi_{i22}(t_f,\, t)\mathbf{p}_i(t) + \int_t^{t_f} [\Phi_{i21}(t_f,\, \tau)\,\mathbf{z}_i(\tau)$$
$$+ \Phi_{i22}(t_f,\, \tau)\,\mathbf{g}_i(\tau)]\, d\tau = \mathbf{0}$$

$$\text{Or} \quad \mathbf{p}_i(t) = -\Phi_{i22}^{-1}(t_f,\, t)\Phi_{i21}(t_f,\, t)\mathbf{x}_i(t) + \Phi_{i22}^{-1}(t_f,\, t)\mathbf{w}_i(t) \tag{10.32}$$

$$\text{where} \quad \mathbf{w}_i(t) = \int_{t_f}^t [\Phi_{i21}(t_f,\, \tau)\mathbf{z}_i(\tau) + \Phi_{i22}(t_f,\, \tau)\mathbf{g}_i(\tau)]\, d\tau \tag{10.33}$$

10.3.2 Riccati matrix and open-loop compensation vector

Upon comparing Eqs. (10.7) and (10.32), we write

$$K_i(t) = -\Phi_{i22}^{-1}(t_f,\, t)\,\Phi_{i21}(t_f,\, t) \tag{10.34}$$
$$\mathbf{s}_i(t) = \Phi_{i22}^{-1}(t_f,\, t)\,\mathbf{w}_i(t) \tag{10.35}$$

From Eq. (10.34), the jth block-pulse coefficient of $K_i(t)$ is given by

$$K_{ij} = -\Phi_{ij\,22}^{-1}\,\Phi_{ij\,21}, \quad j = 0,\, 1,\, 2,\, \ldots,\, \mu - 1. \tag{10.36}$$

Let the BPF representations of $\mathbf{w}_i(t)$, $\Phi_{i21}(t_f,\, t)\,\mathbf{z}_i(t)$ and $\Phi_{i22}(t_f,\, t)\mathbf{g}_i(t)$ be

$$\mathbf{w}_i(t) \simeq \sum_{j=0}^{\mu-1} \mathbf{w}_{ij} B_j(t) = W_i \mathbf{B}(t) \tag{10.37}$$

$$\Phi_{i21}(t_f,\, t)\,\mathbf{z}_i(t) \simeq \sum_{j=0}^{\mu-1} \tilde{\mathbf{z}}_{ij} B_j(t) = \tilde{Z}_i \mathbf{B}(t) \tag{10.38}$$

$$\Phi_{i22}(t_f,\, t)\mathbf{g}_i(t) \simeq \sum_{j=0}^{\mu-1} \tilde{\mathbf{g}}_{ij} B_j(t) = \tilde{G}_i \mathbf{B}(t) \tag{10.39}$$

where

$$W_i = \begin{bmatrix} \mathbf{w}_{i0}, & \mathbf{w}_{i1}, & \cdots, & \mathbf{w}_{i,\,\mu-1} \end{bmatrix} \tag{10.40}$$

$$\tilde{Z}_i = \begin{bmatrix} \Phi_{i021}\mathbf{z}_{i0}, & \Phi_{i121}\mathbf{z}_{i1}, & \cdots, & \Phi_{i,\,\mu-1,21}\mathbf{z}_{i,\,\mu-1} \end{bmatrix} \tag{10.41}$$

$$\tilde{G}_i = \begin{bmatrix} \Phi_{i022}\mathbf{g}_{i0}, & \Phi_{i122}\mathbf{g}_{i1}, & \cdots, & \Phi_{i,\,\mu-1,22}\mathbf{g}_{i,\,\mu-1} \end{bmatrix} \tag{10.42}$$

Substituting Eqs. (10.37)–(10.39) into Eq. (10.33), using the backward integration operational property in Eq. (2.12), and simplifying, we obtain the following recursive relations to compute $\mathbf{w}_i(t)$:

$$\mathbf{w}_{i,\,\mu-1} = -0.5T\left(\Phi_{i,\,\mu-1,21}\mathbf{z}_{i,\,\mu-1} + \Phi_{i,\,\mu-1,22}\mathbf{g}_{i,\,\mu-1}\right) \tag{10.43}$$

$$\begin{aligned} \mathbf{w}_{i,j} = \mathbf{w}_{i,j+1} - 0.5T\left[(\Phi_{i,j,21}\mathbf{z}_{i,j} + \Phi_{i,j,22}\mathbf{g}_{i,j})\right. \\ \left. + (\Phi_{i,j+1,21}\mathbf{z}_{i,j+1} + \Phi_{i,j+1,22}\mathbf{g}_{i,j+1})\right] \end{aligned} \tag{10.44}$$

for $j = \mu - 2,\ \mu - 3,\ \ldots,\ 1,\ 0$.

Now the jth block-pulse coefficient of $\mathbf{s}_i(t)$ can be written from Eq. (10.35) as

$$\mathbf{s}_{i,j} = \Phi_{i,j,22}^{-1}\mathbf{w}_{i,j} \quad \text{for } j = 0,\ 1,\ 2,\ \ldots,\ \mu - 1. \tag{10.45}$$

10.3.3 State vector

Since

$$\dot{\mathbf{x}}_i(t) = \left(A_i - B_i R_i^{-1} B_i^T K_i(t)\right)\mathbf{x}_i(t) - B_i R_i^{-1} B_i^T \mathbf{s}_i(t) + \mathbf{z}_i(t) \tag{10.46}$$

integrating Eq. (10.46) with respect to time t, we have

$$\begin{aligned} \mathbf{x}_i(t) - \mathbf{x}_i(t_0) = \int_{t_0}^t \left[\left(A_i - B_i R_i^{-1} B_i^T K_i(\tau)\right)\mathbf{x}_i(\tau)\right. \\ \left. - B_i R_i^{-1} B_i^T \mathbf{s}_i(\tau) + \mathbf{z}_i(\tau)\right] d\tau \end{aligned} \tag{10.47}$$

Substituting the BPF representation of $\mathbf{s}_i(t)$,

$$\mathbf{x}_i(t) \simeq \sum_{j=0}^{\mu-1} \mathbf{x}_{ij} B_j(t) = X_i \boldsymbol{B}(t) \tag{10.48}$$

$$\mathbf{x}_i(t_0) = \mathbf{x}_i(t_0) \sum_{j=0}^{\mu-1} B_j(t) = X_i(t_0)\boldsymbol{B}(t) \tag{10.49}$$

$$K_i(t)\mathbf{x}_i(t) \simeq \sum_{j=0}^{\mu-1} K_{ij}\mathbf{x}_{ij} B_j(t) = \tilde{X}_i \boldsymbol{B}(t) \tag{10.50}$$

$$\mathbf{z}_i(t) = \sum_{j=0}^{\mu-1} \mathbf{z}_{ij} B_j(t) = Z_i \boldsymbol{B}(t) \tag{10.51}$$

where

$$\tilde{X}_i = \left[\ K_{i0}\mathbf{x}_{i0},\ K_{i1}\mathbf{x}_{i1},\ \ldots,\ K_{i,\mu-1}\mathbf{x}_{i,\mu-1}\ \right], \tag{10.52}$$

and employing forward integration operational property in Eq. (2.10), we obtain

$$X_i - X_i(t_0) = \left[A_i X_i - B_i R_i^{-1} B_i^T \tilde{X}_i - B_i R_i^{-1} B_i^T S_i + Z_i\right] H_f \tag{10.53}$$

Substituting H_f in Eq. (2.11) and simplifying lead to the following recursive relations :

$$\mathbf{x}_{i0} = M_{i0}^{-1}\left[\mathbf{x}_i(t_0) + \mathbf{v}_{i0}\right] \tag{10.54}$$

$$\mathbf{x}_{ij} = M_{ij}^{-1}\left[N_{i,j-1}\mathbf{x}_{i,j-1} + \mathbf{v}_{i,j-i} + \mathbf{v}_{ij}\right] \tag{10.55}$$

$$\text{for } j = 1, 2, \ldots, \mu - 1.$$

where

$$M_{ij} = \left[I_{n_i} - 0.5T\left(A_i - B_i R_i^{-1} B_i^T K_{ij}\right)\right] \tag{10.56}$$

$$N_{ij} = \left[I_{n_i} + 0.5T\left(A_i - B_i R_i^{-1} B_i^T K_{ij}\right)\right] \tag{10.57}$$

$$\mathbf{v}_{ij} = 0.5T\left(\mathbf{z}_{ij} - B_i R_i^{-1} B_i^T \mathbf{s}_{ij}\right) \tag{10.58}$$

10.3.4 Adjoint vector and local control

The jth block-pulse coefficients of $\mathbf{p}_i(t)$ and $\mathbf{u}_i(t)$ can be written from Eqs. (10.6) and (10.7) as

$$\mathbf{p}_{ij} = K_{ij}\mathbf{x}_{ij} + \mathbf{s}_{ij} \tag{10.59}$$

$$\mathbf{u}_{ij} = -R_i^{-1}B_i^T\mathbf{p}_{ij} \tag{10.60}$$

$$\text{for } j = 0, 1, 2, \ldots, \mu - 1.$$

10.3.5 Coordination

The jth block-pulse coefficients of $\boldsymbol{\lambda}_i(t)$ and $\mathbf{z}_i(t)$ can be computed from Eq. (10.17) as

$$\boldsymbol{\lambda}_{ij}^{k+1} = -\mathbf{p}_{ij}^k \tag{10.61}$$

$$\mathbf{z}_{ij}^{k+1} = \left(\sum_{\substack{l=1 \\ l \neq i}}^{N} L_{il}\mathbf{x}_{lj}\right)^k \tag{10.62}$$

10.3.6 Error

Error

$$= \left\{\frac{1}{\mu}\sum_{i=1}^{N}\sum_{j=0}^{\mu-1}\left[\mathbf{z}_{ij}^{k+1} - \left(\sum_{\substack{l=1 \\ l \neq i}}^{N} L_{il}\mathbf{x}_{lj}\right)^k\right]^T\left[\mathbf{z}_{ij}^{k+1} - \left(\sum_{\substack{l=1 \\ l \neq i}}^{N} L_{il}\mathbf{x}_{lj}\right)^k\right]\right\}^{1/2} \tag{10.63}$$

10.4 Extension to Linear Time-Varying Systems

Since A_i, B_i, Q_i and R_i are time-varying, we have

$$F_i(t) = \begin{bmatrix} A_i(t) & -B_i(t)R_i^{-1}(t)B_i^T(t) \\ -Q_i(t) & -A_i^T(t) \end{bmatrix} \tag{10.64}$$

Then

$$\begin{bmatrix} \dot{\mathbf{x}}_i \\ \dot{\mathbf{p}}_i \end{bmatrix} = F_i(t) \begin{bmatrix} \mathbf{x}_i \\ \mathbf{p}_i \end{bmatrix} + \begin{bmatrix} \mathbf{z}_i \\ \mathbf{g}_i \end{bmatrix} \qquad (10.65)$$

Eq. (10.21) becomes

$$\dot{\Phi}_i(t_f, t) = -\Phi_i(t_f, t)F_i(t) \qquad (10.66)$$

Integrating backward with respect to time t gives

$$\Phi_i(t_f, t) - I_{2n_i} = -\int_{t_f}^{t} \Phi_i(t_f, \tau)F_i(\tau)\, d\tau \qquad (10.67)$$

Substituting Eqs. (10.23) and (10.25) into Eq. (10.67), and making use of the disjoint property in Eq. (2.14) and the backward integration operational property in Eq. (2.12) lead to Eq. (10.27) where

$$\hat{\Phi}_i = \begin{bmatrix} \Phi_{i0}F_{i0}, & \Phi_{i1}F_{i1}, & \dots, & \Phi_{i,\,\mu-1}F_{i,\,\mu-1} \end{bmatrix} \qquad (10.68)$$

Upon substituting H_b in Eq. (2.13) into Eq. (10.27) and simplifying, we obtain

$$\Phi_{i,\,\mu-1} = (I_{2n_i} - 0.5TF_{i,\,\mu-1})^{-1} \qquad (10.69)$$

$$\Phi_{i,\,j} = \Phi_{i,\,j+1}(I_{2n_i} + 0.5TF_{i,j+1})(I_{2n_i} - 0.5TF_{i,j})^{-1} \quad (10.70)$$

$$\text{for} \quad j = \mu - 2, \mu - 3, \dots, 1, 0.$$

Next, integrating

$$\dot{\mathbf{x}}_i(t) = \left(A_i(t) - B_i(t)R_i^{-1}(t)B_i^T(t)K_i(t)\right)\mathbf{x}_i(t)$$
$$- B_i(t)R_i^{-1}(t)B_i^T(t)\mathbf{s}_i(t) + \mathbf{z}_i(t) \qquad (10.71)$$

with respect to time t, we have

$$\mathbf{x}_i(t) - \mathbf{x}_i(t_0) = \int_{t_0}^{t}\left[\left(A_i(\tau) - B_i(\tau)R_i^{-1}(\tau)B_i^T(\tau)K_i(\tau)\right)\mathbf{x}_i(\tau)\right.$$
$$\left. - B_i(\tau)R_i^{-1}(\tau)B_i^T(\tau)\mathbf{s}_i(\tau) + \mathbf{z}_i(\tau)\right] d\tau \quad (10.72)$$

Substituting the BPF representations of $\mathbf{s}_i(t)$, $\mathbf{x}_i(t)$, $\mathbf{x}_i(t_0)$, $\mathbf{K}_i(t)$ and $\mathbf{z}_i(t)$ using the disjoint property of BPFs, and employing the forward integration operational property, we have

$$X_i - X_i(t_0) \;=\; [X_{ai} - X_{bi} - S_{bi} + Z_i]\, H_f \qquad (10.73)$$

where

$$X_{ai} \;=\; [\; A_{i0}\mathbf{x}_{i0}, \;\; A_{i1}\mathbf{x}_{i1}, \;\; \ldots, \;\; A_{i,\,\mu-1}\mathbf{x}_{i,\,\mu-1} \;] \qquad (10.74)$$

$$X_{bi} \;=\; [\; B_{i0}R_{i0}^{-1}B_{i0}^{T}K_{i0}\mathbf{x}_{i0}, \;\; B_{i1}R_{i1}^{-1}B_{i1}^{T}K_{i1}\mathbf{x}_{i1}, \;\; \ldots\ldots$$
$$\ldots\ldots, \;\; B_{i,\,\mu-1}R_{i\,\mu-1}^{-1}B_{i\,\mu-1}^{T}K_{i\,\mu-1}\mathbf{x}_{i,\,\mu-1} \;] \qquad (10.75)$$

$$S_{bi} \;=\; [\; B_{i0}R_{i0}^{-1}B_{i0}^{T}\mathbf{s}_{i0}, \;\; B_{i1}R_{i1}^{-1}B_{i1}^{T}\mathbf{s}_{i1}, \;\; \ldots,$$
$$B_{i,\,\mu-1}R_{i\,\mu-1}^{-1}B_{i\,\mu-1}^{T}\mathbf{s}_{i,\,\mu-1}] \qquad (10.76)$$

Substituting H_f and simplifying lead to the following recursive relations :

$$\mathbf{x}_{i0} \;=\; \tilde{M}_{i0}^{-1}\left[\mathbf{x}_i(t_0) + \tilde{\mathbf{v}}_{i0}\right] \qquad (10.77)$$

$$\mathbf{x}_{ij} \;=\; \tilde{M}_{ij}^{-1}\left[\tilde{N}_{i,j-1}\,\mathbf{x}_{i,j-1} + \tilde{\mathbf{v}}_{i,j-i} + \tilde{\mathbf{v}}_{ij}\right] \qquad (10.78)$$

$$\text{for } j \;=\; 1,\, 2,\, \ldots,\, \mu - 1.$$

where

$$\tilde{M}_{ij} \;=\; \left[I_{n_i} - 0.5T\left(A_{ij} - B_{ij}R_{ij}^{-1}B_{ij}^{T}K_{ij}\right)\right] \qquad (10.79)$$

$$\tilde{N}_{ij} \;=\; \left[I_{n_i} + 0.5T\left(A_{ij} - B_{ij}R_{ij}^{-1}B_{ij}^{T}K_{ij}\right)\right] \qquad (10.80)$$

$$\tilde{\mathbf{v}}_{ij} \;=\; 0.5T\left(\mathbf{z}_{ij} - B_{ij}R_{ij}^{-1}B_{ij}^{T}\mathbf{s}_{ij}\right) \qquad (10.81)$$

The jth block-pulse coefficient of $\mathbf{u}_i(t)$ can be written as

$$\mathbf{u}_{ij} \;=\; -R_{ij}^{-1}B_{ij}^{T}\mathbf{p}_{ij} \quad \text{for } j \;=\; 0,\, 1,\, 2,\, \ldots,\, \mu - 1. \quad (10.82)$$

10.5 Computational Algorithm

The computational algorithm is

step 1: Evaluate $\Phi_i(t_f, t)$ and $K_i(t)$ from Eqs. (10.29), (10.30)(or from Eqs. (10.69), (10.70)) and (10.36) for $i = 1, 2, \ldots, N$ and store $K_i(t)$. Initialize an arbitrary value for $\lambda_i(t)$ and find $z_i(t)$. Set $k = 1$.

step 2: At the kth iteration, use the values of $\lambda_i^k(t)$ and $z_i^k(t)$ to evaluate $s_i(t)$ from Eqs. (10.43)−(10.45) and store $s_i(t)$ for $i = 1, 2, \ldots, N$.

step 3: Evaluate $x_i(t)$ from Eqs. (10.54)−(10.58) (or from Eqs. (10.77)−(10.81)) and store for $i = 1, 2, \ldots, N$.

step 4: Evaluate $p_i(t)$ from Eq. (10.59) and store for $i = 1, 2, \ldots, N$.

step 5: At the second level, use the results of steps 2−4 and Eqs. (10.61) and (10.62) to update the coordination vector.

step 6: Check for the convergence at the second level by evaluating the over all interaction (communication) error using Eq. (10.63).

step 7: If the desired convergence is achieved, stop. Else, set $k = k+1$ and go to step 2.

10.6 Illustrative Examples

Example 1

Consider a two-reach model of a river pollution control problem [21]

$$
\begin{bmatrix} \dot{x}_1 \\ \dot{x}_2 \\ \cdots \\ \dot{x}_3 \\ \dot{x}_4 \end{bmatrix} = \begin{bmatrix} -1.32 & 0 & \vdots & 0 & 0 \\ -0.32 & -1.2 & \vdots & 0 & 0 \\ \cdots & \cdots & \cdots & \cdots & \cdots \\ 0.90 & 0 & \vdots & -1.32 & 0 \\ 0 & 0.9 & \vdots & -0.32 & -1.2 \end{bmatrix} \begin{bmatrix} x_1 \\ x_2 \\ \cdots \\ x_3 \\ x_4 \end{bmatrix}
$$
$$
+ \begin{bmatrix} 0.1 & \vdots & 0 \\ 0 & \vdots & 0 \\ \cdots & \cdots & \cdots \\ 0 & \vdots & 0.1 \\ 0 & \vdots & 0 \end{bmatrix} \begin{bmatrix} u_1 \\ \cdots \\ u_2 \end{bmatrix}
$$

and the quadratic cost function

$$
J = \frac{1}{2} \int_0^5 \left[\mathbf{x}^T Q \mathbf{x} + \mathbf{u}^T R \mathbf{u} \right] dt
$$

with the initial state vector

$$
\begin{bmatrix} x_1(0) \\ x_2(0) \\ x_3(0) \\ x_4(0) \end{bmatrix} = \begin{bmatrix} 1 \\ 1 \\ -1 \\ 1 \end{bmatrix}, \quad Q = \begin{bmatrix} 2 & 0 & 0 & 0 \\ 0 & 4 & 0 & 0 \\ 0 & 0 & 2 & 0 \\ 0 & 0 & 0 & 4 \end{bmatrix}, \quad \text{and } R = \begin{bmatrix} 2 & 0 \\ 0 & 2 \end{bmatrix}
$$

This system is decomposed into two subsystems where each subsystem has two states. The interaction variables are $z_1 = x_1$ and $z_2 = x_2$. This problem is solved with 20 BPFs and error $= 10^{-5}$. The control and state trajectories are shown in Figs 10.2 and 10.3. Fig. 10.4 shows how the error decreases to zero approximately with only three iterations.

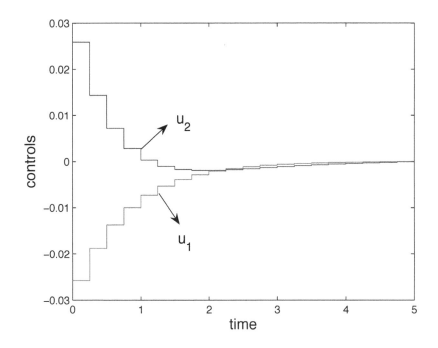

Figure 10.2: Control variables u_1 and u_2.

Example 2

Consider a four-system large scale system (Pearson's 12th order example) shown in Fig. 10.5. The solid lines represent inputs and outputs of the systems while the dashed lines represent data transmission among systems. The dynamic equations of this large scale system are given [13] as follows :

System 1 :

$$\begin{bmatrix} \dot{x}_1 \\ \dot{x}_2 \\ \dot{x}_3 \end{bmatrix} = \begin{bmatrix} 0 & 1 & 0 \\ 0 & 0 & 1 \\ -1 & -2 & -3 \end{bmatrix} \begin{bmatrix} x_1 \\ x_2 \\ x_3 \end{bmatrix} + \begin{bmatrix} 0 \\ 0 \\ 1 \end{bmatrix} u_1 + \begin{bmatrix} 0 \\ 0 \\ 1 \end{bmatrix} z_1$$

$$\begin{bmatrix} x_1(0) \\ x_2(0) \\ x_3(0) \end{bmatrix} = \begin{bmatrix} -1.2 \\ 0.2 \\ -1.0 \end{bmatrix} ; \quad \begin{bmatrix} y_1 \\ y_2 \end{bmatrix} = \begin{bmatrix} 1 & 0 & 0 \\ 0 & 1 & 0 \end{bmatrix} \begin{bmatrix} x_1 \\ x_2 \\ x_3 \end{bmatrix}$$

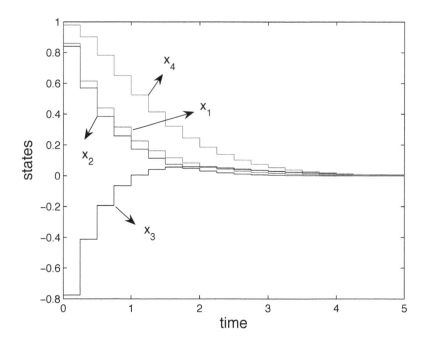

Figure 10.3: State variables x_1, x_2, x_3 and x_4.

and $z_1 = x_{11}$ where x_{11} is a state of system 4.

System 2 :

$$
\begin{bmatrix} \dot{x}_4 \\ \dot{x}_5 \\ \dot{x}_6 \end{bmatrix} = \begin{bmatrix} 0 & 1 & 0 \\ 0 & 0 & 1 \\ -2 & -3 & -1 \end{bmatrix} \begin{bmatrix} x_4 \\ x_5 \\ x_6 \end{bmatrix} + \begin{bmatrix} 0 & 0 \\ 0 & 0 \\ 1 & 1 \end{bmatrix} \begin{bmatrix} z_2 \\ z_3 \end{bmatrix}
$$

$$
\begin{bmatrix} x_4(0) \\ x_5(0) \\ x_6(0) \end{bmatrix} = \begin{bmatrix} 0.4 \\ -0.8 \\ 0.6 \end{bmatrix} ; \quad \begin{bmatrix} y_3 \\ y_4 \end{bmatrix} = \begin{bmatrix} 1 & 0 & 0 \\ 0 & 1 & 0 \end{bmatrix} \begin{bmatrix} x_4 \\ x_5 \\ x_6 \end{bmatrix}
$$

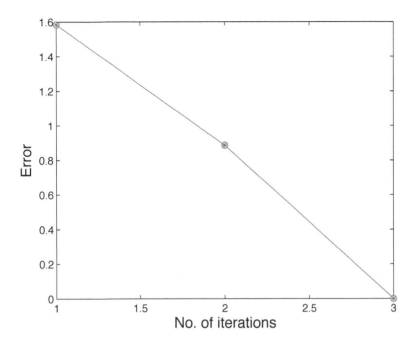

Figure 10.4: Interaction error behaviour.

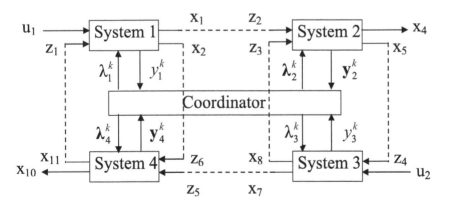

Figure 10.5: A large scale system of four systems and its hierarchical control.

where $z_2 = x_1$ and $z_3 = x_8$.

System 3 :

$$\begin{bmatrix} \dot{x}_7 \\ \dot{x}_8 \\ \dot{x}_9 \end{bmatrix} = \begin{bmatrix} 0 & 1 & 0 \\ 0 & 0 & 1 \\ -3 & -1 & -2 \end{bmatrix} \begin{bmatrix} x_7 \\ x_8 \\ x_9 \end{bmatrix} + \begin{bmatrix} 0 \\ 0 \\ 1 \end{bmatrix} u_2 + \begin{bmatrix} 0 \\ 0 \\ 1 \end{bmatrix} z_4$$

$$\begin{bmatrix} x_7(0) \\ x_8(0) \\ x_9(0) \end{bmatrix} = \begin{bmatrix} -0.6 \\ 0.4 \\ -0.4 \end{bmatrix} ; \quad \begin{bmatrix} y_5 \\ y_6 \end{bmatrix} = \begin{bmatrix} 1 & 0 & 0 \\ 0 & 1 & 0 \end{bmatrix} \begin{bmatrix} x_7 \\ x_8 \\ x_9 \end{bmatrix}$$

and $z_4 = x_5$.

System 4 :

$$\begin{bmatrix} \dot{x}_{10} \\ \dot{x}_{11} \\ \dot{x}_{12} \end{bmatrix} = \begin{bmatrix} 0 & 1 & 0 \\ 0 & 0 & 1 \\ -1 & -2 & -3 \end{bmatrix} \begin{bmatrix} x_{10} \\ x_{11} \\ x_{12} \end{bmatrix} + \begin{bmatrix} 0 & 0 \\ 0 & 0 \\ 1 & 1 \end{bmatrix} \begin{bmatrix} z_5 \\ z_6 \end{bmatrix}$$

$$\begin{bmatrix} x_{10}(0) \\ x_{11}(0) \\ x_{12}(0) \end{bmatrix} = \begin{bmatrix} 1.0 \\ -0.2 \\ 1.2 \end{bmatrix} ; \quad \begin{bmatrix} y_7 \\ y_8 \end{bmatrix} = \begin{bmatrix} 1 & 0 & 0 \\ 0 & 1 & 0 \end{bmatrix} \begin{bmatrix} x_{10} \\ x_{11} \\ x_{12} \end{bmatrix}$$

where $z_5 = x_7$ and $z_6 = x_2$.

The system and control weighting matrices are chosen to be

$$Q_i = \text{diag}[1, 1, 0], \quad R_i = 1 \quad \text{for} \quad i = 1, 2, 3, 4.$$

and t_0 and t_f are chosen to be 0 and 15 respectively.

This problem is solved with 60 BPFs and error 0.01. The computed control inputs $u_1(t)$, $u_2(t)$; state variables $x_1(t)$, $x_2(t)$, ..., $x_{12}(t)$; and the interaction error behaviour are shown in Figs. 10.6 - 10.11. It is evident from these figures that the results are satisfactory.

Example 3

Consider the problem [45] of minimizing

$$J = \frac{1}{2} \int_0^1 \left[\mathbf{x}^T(t) Q \mathbf{x}(t) + \mathbf{u}^T(t) R \mathbf{u}(t) \right] dt$$

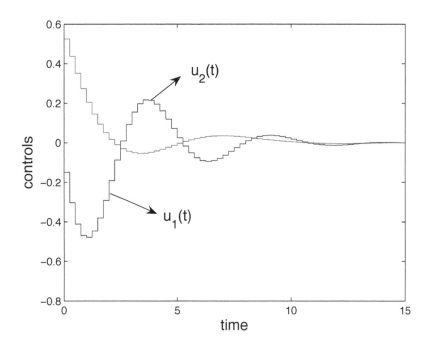

Figure 10.6: Control variables u_1 and u_2.

subject to

$$\left[\begin{array}{c} \dot{x}_1(t) \\ \dot{x}_2(t) \end{array}\right] = \left[\begin{array}{cc} -10t & 1 \\ 1 & -5t^2 \end{array}\right] \left[\begin{array}{c} x_1(t) \\ x_2(t) \end{array}\right] + \left[\begin{array}{c} u_1(t) \\ u_2(t) \end{array}\right]$$

with

$$\left[\begin{array}{c} x_1(0) \\ x_2(0) \end{array}\right] = \left[\begin{array}{c} 5 \\ 10 \end{array}\right], \quad Q = 10\left[\begin{array}{cc} 1 & 0 \\ 0 & 1 \end{array}\right], \quad R = \left[\begin{array}{cc} 1 & 0 \\ 0 & 1 \end{array}\right]$$

This system is decomposed into two subsystems with the interaction variables $z_1 = x_2$ and $z_2 = x_1$ and is solved with 20 BPFs and error $= 10^{-6}$ The control and state trajectories are shown in Figs 10.12 and 10.13. Fig. 10.14 shows how the error decreases with the increasing iteration number.

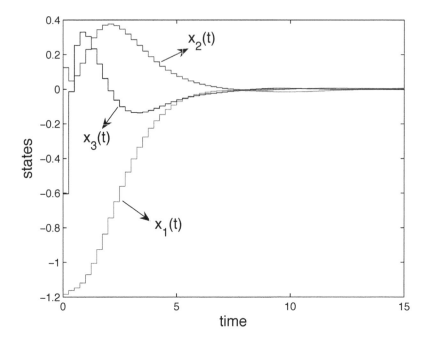

Figure 10.7: State variables x_1, x_2 and x_3.

10.7 Conclusion

A novel hierarchical control paradigm has been suggested for controlling linear time-invariant/time-varying large scale systems via BPFs. Compared with the conventional methods reported in the literature [13, 21], the BPF method is computationally simpler and attractive as it is totally recursive in nature. In the conventional method one has to solve matrix Riccati (nonlinear) differential equations using some iterative technique. In the suggested method we follow state transition matrix approach and compute everything in a recursive manner.

Notice from Figs. 10.2, 10.3, 10.6–10.10, 10.12 and 10.13 that all the control variables and state variables are piecewise con-

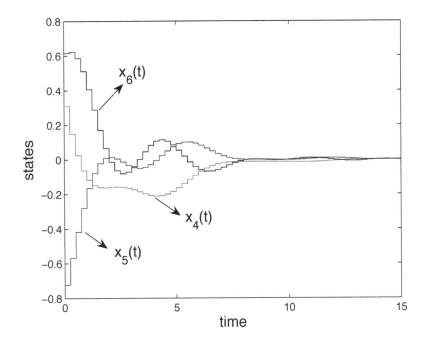

Figure 10.8: State variables x_4, x_5 and x_6.

stant, i.e. the solution is continuous. This is not the case with
the conventional numerical methods which always produce a discrete solution. As BPFs are piecewise constant basis functions,
the computed result is inherently piecewise constant. Each pulse
in the solution corresponds to average value of the signal over the
block-pulse width, i.e. T. One may argue that a piecewise constant
(continuous) signal can be obtained by passing the discrete data,
obtained from the conventional method, through a zero-order hold.
But, each pulse in this case does not represent the average value
of the signal over T.

Since we are able to generate a piecewise constant signal directly from the BPF solution, which is just a sequence of numbers
with each number representing a pulse in the solution, the BPF

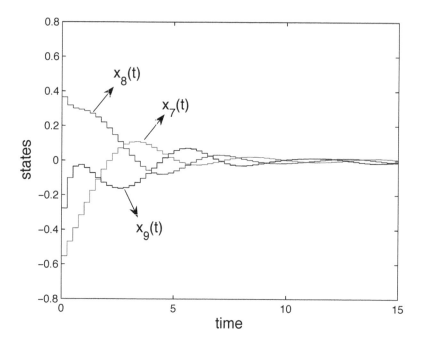

Figure 10.9: State variables x_7, x_8 and x_9.

approach is quite suitable for digital control of continuous-time systems. Moreover, it is completely recursive, making its application very attractive in dealing with computations of large-scale systems.

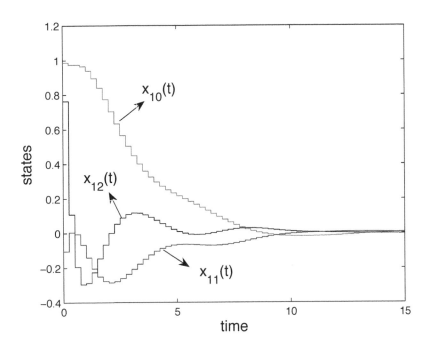

Figure 10.10: State variables x_{10}, x_{11} and x_{12}.

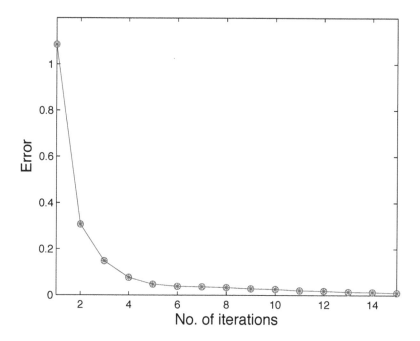

Figure 10.11: Interaction error behaviour.

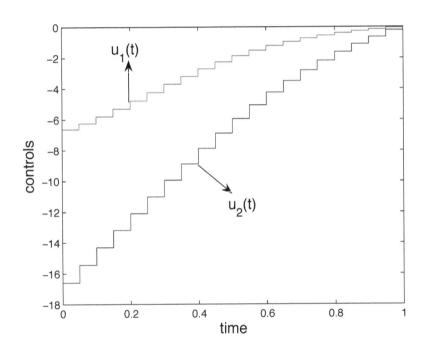

Figure 10.12: Control variables u_1 and u_2.

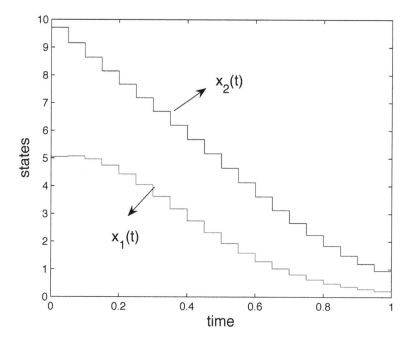

Figure 10.13: State variables x_1 and x_2.

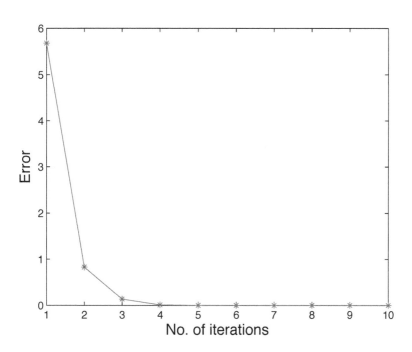

Figure 10.14: Interaction error behaviour.

Chapter 11

Epilogue

In general, the optimal control of closed-loop systems requires the solution of a set of nonlinear differential equations [3, 10]. Since OFs are used, in the present approach all that needs to be solved is a set of linear/non-linear algebraic equations, and the accuracy of the solution can be improved to the desired degree by increasing the number of OFs.

It is always tempting to propose a new algorithm; design and optimize its parameters to get a more effective and simple algorithm; test the performance with speed and accuracy in computation and finally compare it with the existing results. In this book, some attempts have been made to solve the optimal control problem of various types of systems having quadratic performance index via OFs. Mostly BPFs and SLPs are used to solve the problems by developing new algorithms. Recursive algorithms are developed wherever possible.

The results presented in this book may encourage others to investigate the following important problems:

- Constrained optimal control

- Hierarchical control of large scale systems

- Optimal control of distributed parameter systems

As a matter of fact, the constrained optimal control problem has already been studied via hybrid functions [66], HFs [69, 88] and BPFs [75]. As these approaches are nonrecursive in nature, there is a need to explore the possibility of developing recursive algorithms yet. Moreover, a further study on constrained optimal control via the OF approach is needed as some of the approaches/results reported in the literature are not convincing.

The hierarchical control problem of large scale systems has also been studied via BPFs [45]. An attempt has been made to solve this problem using SLPs but failed. It is observed that solving this problem via SLPs, in a manner similar to BPF approach, is not easy because SLPs lack disjoint property.

Optimal control of distributed parameter systems has also been studied via SLPs [17, 60], SCP1s [29, 63], and BPFs [50, 51] with some success. The approaches using SLPs and SCP1s were found to be nonrecursive. The BPF approach in [50, 51] appears to be incorrect for the following important reason:

The approach makes use of differentiation operational matrix of BPFs. Since BPFs are constant disjoint functions their differentiation operational matrix can't be found by direct differentiation of BPFs. It is found in an indirect manner as explained here.

Let

$$\frac{d}{dt} \mathbf{B}(t) = \mathbb{D}\mathbf{B}(t)$$

where $\mathbb{D}$ denotes differentiation operational matrix of order $m \times m$. Integrating the above equation with respect to t once, we have

$$\boldsymbol{B}(t) - \boldsymbol{B}(0) = \mathbb{D} \int_0^t \boldsymbol{B}(\tau)d\tau$$

$$\text{or} \quad \left[I - \boldsymbol{B}(0)\mathbf{u}^T \right] \boldsymbol{B}(t) = \mathbb{D} H_f \boldsymbol{B}(t)$$

Hence

$$\mathbb{D} = \left[I - \boldsymbol{B}(0)\mathbf{u}^T \right] H_f^{-1} = \mathbb{D}_1, \quad \text{say}$$

where $\mathbf{u} = \left[1 \;\; 1 \;\; \ldots \;\; 1 \right]^T$, an m - vector, and H_f is the forward integration operational matrix of order $m \times m$ which is shown in Eq. (2.11). But, according to Zhu and Lu [50, 51] $\mathbb{D}$ is given by

$$\mathbb{D} = H_f^{-1} \left[I - \boldsymbol{B}(0)\mathbf{u}^T \right] = \mathbb{D}_2, \quad \text{say}$$

which is incorrect.

Let us now see what results $\mathbb{D}_1$ and $\mathbb{D}_2$ produce by considering an example.

Let $f(t) = 1 + t$ over $0 \le t \le 4$. Then

$$g(t) = \frac{d}{dt} f(t) = 1$$

The block-pulse spectra of $f(t)$ and $g(t)$, with $m = 4$, are given by

$$\mathbf{f} = \left[1.5, \;\; 2.5, \;\; 3.5, \;\; 4.5 \right]^T$$

$$\text{and} \quad \mathbf{g} = \left[1, \;\; 1, \;\; 1, \;\; 1 \right]^T$$

respectively. Since

$$H_f^{-1} = 4 \begin{bmatrix} 0.5 & -1 & 1 & -1 \\ 0 & 0.5 & -1 & 1 \\ 0 & 0 & 0.5 & -1 \\ 0 & 0 & 0 & 0.5 \end{bmatrix}$$

$\mathbf{g}$ via $\mathbb{D}_1$ is given by

$$\mathbf{g}^T = \mathbf{f}^T \mathbb{D}_1 = \begin{bmatrix} 0, & 2, & 0, & 2 \end{bmatrix} = \mathbf{g}_1^T, \quad \text{say}$$

and $\mathbf{g}$ via $\mathbb{D}_2$ is given by

$$\mathbf{g}^T = \mathbf{f}^T \mathbb{D}_2 = \begin{bmatrix} 0, & -4, & 0, & -4 \end{bmatrix} = \mathbf{g}_2^T, \quad \text{say}$$

By comparing $\mathbf{g}_1$ and $\mathbf{g}_2$ with $\mathbf{g}$ we observe that

- the elements of $\mathbf{g}_1$ and $\mathbf{g}_2$ are nowhere near the elements of $\mathbf{g}$,

- the elements of $\mathbf{g}_1$ are nonnegative while the elements of $\mathbf{g}$ are all positive, and

- the elements of $\mathbf{g}_2$ are nonpositive while the elements of $\mathbf{g}$ are all positive.

In view of the above observations one can say that both $\mathbb{D}_1$ and $\mathbb{D}_2$ are incorrect, though the manner in which $\mathbb{D}_1$ is found appears to be logical and correct. So, what is $\mathbb{D}$? Unless the issue of finding the correct $\mathbb{D}$ is settled, the optimal control problem of distributed parameter systems cannot be solved. Moreover, one has to be cautious while applying OF approach to study the problems of distributed parameter systems as it may become numerically unstable for systems of order two or more than two. It has already been demonstrated [22] that the BPF approach becomes numerically unstable for second order systems. The authors also have noticed numerical instabilities while working with other classes of OFs. So, a deeper study on the application of OFs to distributed parameter systems is very much necessary.

Bibliography

[1] D. Luenberger, "Observing the state of a linear system," *IEEE Trans. on Military Electronics*, vol. 8, pp: 74-80, 1964.

[2] J. J. Bongiorno, Jr. and D. C. Youla, "On observers in multivariable control systems," *Int. J. Control*, vol. 8, no. 3, pp: 221-243, 1968.

[3] M. Athans and P. L. Falb, *Optimal Control*, Lincoln Laboratory Publications, 1968.

[4] M. Athans, "The role and use of the stochastic linear-quadratic-Gaussian problem in control system design," *IEEE Trans. Automatic Control*, vol. 16, no. 6, pp: 529-552, 1971.

[5] M. Kline, *Mathematical thoughts from Ancient to Modern Times*, Oxford University Press, New York, 1972.

[6] C. F. Chen and C. H. Hsiao, "Walsh series analysis in optimal control," *Int. J. Control*, vol. 21, no. 6, pp: 881-897, 1975.

[7] K. Ogata, *State Space Analysis of Control Systems*, Prentice-Hall Inc., Englewood Cliffs, N. J., 1976

[8] C. F. Chen, Y. T. Tsay and T. T. Wu, "Walsh operational matrices for fractional calculus and their application to dis-

tributed systems," *J. The Franklin Institute*, vol. 303, no. 3,
pp: 267-284, 1977.

[9] P. Sannuti, "Analysis and synthesis of dynamic systems via
block-pulse functions," *Proc. IEE*, vol. 124, no. 6, pp: 569-571,
1977.

[10] A. P. Sage and C. C. White, *Optimum Systems Control*,
Prentice-Hall, Inc., Englewood Cliffs, New Jersey, 1977.

[11] Y. A. Kochetkov and V. K. Tomshin, "Optimal control of
deterministic systems described by integrodifferential equations," *Automation and Remote Control*, vol. 39, no. 1, pp:
1-6, 1978.

[12] J. W. Brewer, "Kronecker products and matrix calculus in
system theory," *IEEE Trans. Circuits and Systems*, vol. 25,
no. 9, pp: 772-781, 1978.

[13] M. G. Singh and A. Titli, *Systems: Decomposition, Optimization and Control*, Pergamon Press, Oxford, 1978.

[14] G. X. Fang, *Computational Methods of Optimal Control Problems*, Beijing: Science Press, 1979.

[15] L. Pandolfi, "On the regulator problem for linear degenerate
control systems," *J. Optimization Theory and Applications*,
vol. 33, no. 2, pp: 241-254, 1981.

[16] N. S. Hsu and B. Cheng, "Analysis and optimal control of
time-varying linear systems via block-pulse functions," *Int.
J. Control*, vol. 33, no. 6, pp: 1107-1122, 1981.

[17] M. L. Wang and R. Y. Chang, "Optimal control of linear distributed parameter systems by shifted Legendre polynomial functions," *Trans. ASME J. Dynamic Systems, Measurement, and Control*, vol. 105, no. 4, pp: 222-226, 1983.

[18] D. Cobb, "Descriptor variable Systems and optimal state regulation," *IEEE Trans. Automatic Control*, vol. 28, no. 5, pp: 601-611, 1983.

[19] S. Kawaji, "Block-pulse series analysis of linear systems incorporating observers," *Int. J. Control*, vol. 37, no. 5, pp: 1113-1120, 1983.

[20] K. R. Palanisamy and G. P. Rao, "Optimal control of linear systems with delays in state and control via Walsh functions," *Proc. IEE*, pt. D, vol. 130, no. 6, pp: 300-312, 1983.

[21] M. Jamshidi, *Large-Scale Systems: Modelling and Control*, North Holland, New York, 1983.

[22] G. P. Rao, *Piecewise Constant Orthogonal Functions and Their Application to Systems and Control*, LNCIS 55, Springer, Berlin, 1983.

[23] N. K. Sinha and Z. Q. Jie, "State estimation using block-pulse functions," *Int. J. Systems Science*, vol. 15, no. 4, pp: 341-350, 1984.

[24] C. Hwang and Y. P. Shih, "Optimal control of delay systems via block-pulse functions," *J. Optimization Theory and Application*, vol. 45, no. 1, pp: 101-112, 1985.

[25] J. H. Chou and I. R. Horng, "Shifted Chebyshev series analysis of linear optimal control systems incorporating observers," *Int. J. Control*, vol. 41, no. 1, pp: 129-134, 1985.

[26] C. Hwang and M. Y. Chen, "Analysis and parameter identification of time-delay systems via shifted Legendre polynomials," *Int. J. Control*, vol. 41, no. 2, pp: 403-415, 1985.

[27] I. R. Horng and J. H. Chou, "Analysis, parameter estimation and optimal control of time-delay systems via Chebyshev series," *Int. J. Control*, vol. 41, no. 5, pp: 1221-1234, 1985.

[28] C. Hwang and M. Y. Chen, "Analysis and optimal control of time-varying linear systems via shifted Legendre polynomials," *Int. J. Control*, vol. 41, no. 5, pp: 1317-1330, 1985.

[29] I. R. Horng and J. H. Chou, "Application of shifted Chebyshev series to the optimal control of linear distributed parameter systems," *Int. J. Control*, vol. 42, no. 1, pp: 233-241, 1985.

[30] J. H. Chou and I. R. Horng, "Shifted Legendre series analysis of linear optimal control systems incorporating observers," *Int. J. Systems Science*, vol. 16, no. 7, pp: 863-867, 1985.

[31] C. Hwang and M. Y. Chen, "Suboptimal control of linear time-varying multi-delay systems via shifted Legendre polynomials," *Int. J. Systems Science*, vol. 16, no. 12, pp: 1517-1537, 1985.

[32] Y. F. Chang and T. T. Lee, "Applications of general orthogonal polynomials to the optimal control of time-varying linear systems," *Int. J. Control*, vol. 43, no. 4, pp: 1283-1304, 1986.

[33] T. T. Lee, S. C. Tsay and I. R. Horng, "Shifted Jacobi series analysis of linear optimal control systems incorporating observers," *J. The Franklin Institute*, vol. 321, no. 5, pp: 289-298, 1986.

[34] D. H. Shih and F. C. Kung, "Optimal control of deterministic systems via shifted Legendre polynonials," *IEEE Trans. Automatic Control*, vol. AC-31, no. 5, pp: 451-454, 1986.

[35] Y. F. Chang and T. T. Lee, "General orthogonal polynomials approximations of the linear-quadratic-Gaussian control design," *Int. J. Control*, vol. 43, no. 6, pp: 1879-1895, 1986.

[36] M. H. Perng, "Direct approach for the optimal control of linear time-delay systems via shifted Legendre polynomials," *Int. J. Control*, vol. 43, no. 6, pp: 1897-1904, 1986.

[37] C. Hwang, D. H. Shih and F. C. Kung, "Use of block-pulse functions in the optimal control of deterministic systems," *Int. J. Control*, vol. 44, no. 2, pp: 343-349, 1986.

[38] V. Lovass-Nagy, R. Schilling and H. C. Yan, "A note on optimal control of generalized state-space (descripter) systems," *Int. J. Control*, vol. 44, no. 3, pp: 613-624, 1986.

[39] J. H. Chou and I. R. Horng, "State estimation using continous orthogonal funtions," *Int. J. Systems Science*, vol. 17, no. 9, pp: 1261-1267, 1986.

[40] T. T. Lee and Y. F. Chang, "Analysis, parameter estimation and optimal control of nonlinear systems via general orthogonal polynomials," *Int. J. Control*, vol. 44, no. 4, pp: 1089-1102, 1986.

[41] Y. F. Chang and T. T. Lee, "General orthogonal polynomials analysis of linear optimal control systems incorporating observers," *Int. J. Systems Science*, vol. 17, no. 11, pp: 1521-1535, 1986.

[42] D. H. Shih and L. F. Wang, "Optimal control of deterministic systems described by integrodifferential equations," *Int. J. Control*, vol. 44, no. 6, pp: 1737-1745, 1986.

[43] J. H. Chou, "Application of Legendre series to the optimal control of integrodifferential equations," *Int. J. Control*, vol. 45, no. 1, pp: 269-277, 1987.

[44] H. Y. Chung and Y. Y. Sun, "Fourier series analysis of linear optimal control systems incorporating observers," *Int. J. Systems Science*, vol. 18, no. 2, pp: 213-220, 1987.

[45] J. M. Zhu and Y. Z. Lu, "New approach to hierarchical control via block-pulse function transformations," *Int. J. Control*, vol. 46, no. 2, pp: 441-453, 1987.

[46] M. M. Zavarei and M. Jamshidi, *Time-Delay Systems Analysis, Optimization and Applications*, North-Holland Systems and Control Series, vol. 9, Amsterdam, 1987.

[47] C. Y. Yang and C. K. Chen, "Linear optimal control systems by reduced-order observers via orthogonal functions," *Int. J. Systems Science*, vol. 19, no. 1, pp: 23-32, 1988.

[48] S. C. Tsay, I. L. Wu and T. T. Lee, "Optimal control of linear time-delay systems via general orthogonal polynomials," *Int. J. Systems Science*, vol. 19, no. 2, pp: 365-376, 1988.

[49] J. Vlassenbroeck and R. V. Dooren, "A Chebyshev technique for solving nonlinear optimal control problems," *IEEE Trans. Automatic Control*, vol. 33, no. 4, pp: 333-340, 1988.

[50] J. M. Zhu and Y. Z. Lu, "Hierarchical optimal control for distributed parameter systems via block-pulse operator," *Int. J. Control*, vol. 48, no. 2, pp: 685-703, 1988.

[51] J. M. Zhu and Y. Z. Lu, "Application of single step method of block-pulse functions to the optimal control of linear distributed parameter systems," *Int. J. Systems Science*, vol. 19, no. 12, pp: 2459-2472, 1988.

[52] K. R. Palanisamy and K. G. Raghunathan, "Single-term Walsh series analysis of linear optimal control systems incorporating observers," *Int. J. Systems Science*, vol. 20, No. 7, pp: 1149-1155, 1989.

[53] L. Dai, *Singular Control Systems*, Springer-Verlag, New York, 1989.

[54] W. Shienyu, "Convergence of block pulse series approximation solution for optimal control problem," *Int. J. Systems Science*, vol. 21, no. 7, pp: 1355-1368, 1990.

[55] M. Razzaghi and M. Razzaghi, "Fourier series approach for the solution of linear two-point boundary value problems with time-varying coefficients," *Int. J. Systems Science*, vol. 21, no. 9, pp: 1783-1794, 1990.

[56] K. Balchandran and K. Murugesan, "Optimal control of singular systems via single-term Walsh series," *Int. J. Computer Mathematics*, vol. 43, no. 3, pp: 153-159, 1992.

[57] Z. H. Jiang and W. Schaufelberger, *Block-Pulse Functions and Their Applications in Control Systems*, LNCIS 179, Springer, Berlin, 1992.

[58] M. Razzaghi, M. F. Habibi and R. Fayzebakhsh, "Suboptimal control of linear delay systems via Legendre series," *Kybernetica*, vol. 31, no. 5, pp: 509-518, 1995.

[59] K. B. Datta and B. M. Mohan, *Orthogonal Functions in Systems and Control*, Advanced Series in Electrical and Computer Engineering, vol. 9, World Scientific, Singapore, 1995.

[60] M. Razzaghi and M. Habibi, "Application of Legendre series to the control problems governed by linear parabolic equations," *Mathematics and Computers in Simulation*, vol. 42, no. 1, pp: 77-84, 1996.

[61] A. Patra and G. P. Rao, *General Hybrid Orthogonal Functions and Their Applications in Systems and Control*, LNCIS 213, Springer, London, 1996.

[62] M. Razzaghi and M. Shafiee, "Optimal control of singular systems via Legendre series," *Int. J. Computer Mathematics*, vol. 70, no. 2, pp: 241-250, 1998.

[63] S. A. Bianco, I. S. Sadek and M. T. Kambule, "Optimal control of a class of time-delayed distributed systems by orthogonal functions," *J. The Franklin Institute*, vol. 335, no. 8, pp: 1477-1492, 1998.

[64] F. Fahroo and I. M. Ross, "Costate estimation by a Legendre pseudospectral method," *J. Guidance, Control, and Dynamics*, vol. 24, no. 2, pp: 270-277, 2001.

[65] H. Jaddu, "Direct solution of nonlinear optimal control problems using quasilinearization and Chebyshev polynomials," *J. The Franklin Institute*, vol. 339, pp: 479-498, 2002.

[66] H. R. Marzban and M. Razzaghi, "Hybrid functions approach for linearly constrained quadratic optimal problems," *Applied Mathematical Modelling*, vol. 27, no. 6, pp: 471-485, 2003.

[67] H. R. Marzban and M. Razzaghi, "Optimal control of linear delay systems via hybrid of block-pulse and Legendre polynomials," *J. The Franklin Institute*, vol. 341, no. 3, pp: 279-293, 2004.

[68] H. R. Karimi, P. J. Maralani, B. Moshiri and B. Lohmann, "Numerically efficient approximations to the optimal control of linear singularly perturbed systems based on Haar wavelets," *Int. J. Computer Mathematics*, vol. 82, no. 4, pp: 495-507, 2005.

[69] Y. Ordokhani and M. Razzaghi, "Linear quadratic optimal control problems with inequality constraints via rationalized Haar functions," *Dynamics of Continous, Discrete and Impulsive Systems Series B: Applications and Algorithms*, vol. 12, no. 5-6, pp: 761-773, 2005.

[70] B. M. Mohan and S. K. Kar, "Comments on: Optimal control via Fourier series of operational matrix of integration", *IEEE Transactions on Automatic Control*, vol. 50, no. 9, 1466-1467, 2005.

[71] R. Ebrahimi, M. Samavat, M. A. Vali and A. A. Gharavisi, "Application of Fourier series direct method to the optimal

control of singular systems," *ICGST-ACSE Journal*, vol. 7, no. 2, pp: 19-24, 2007.

[72] X. T. Wang, "Numerical solution of optimal control for time-delay systems by hybrid of block-pulse functions and Legendre polynomials," *Applied Mathematics and Computation*, vol. 184, no. 2, pp: 849-856, 2007.

[73] X. T. Wang, "Numerical solution of optimal control for linear time-varying systems with delays via hybrid functions," *J. The Franklin Institute*, vol. 344, no. 7, pp: 941-953, 2007.

[74] S. H. Chen, W. H. Ho and J. H. Chou, "Design of robust quadratic-optimal controllers for uncertain singular systems using orthogonal function approach and genetic algorithm," *Optimal Control Applications and Methods*, vol. 29, no. 5, pp: 373-391, 2008.

[75] N. Boussiala, H. Chaabi and W. Liu, "Numerical methods for solving constrained nonlinear optimal control using the block pulse functions," *Int. J. Innovative Computing, Information and Control*, vol. 4, no. 7, pp: 1733-1740, 2008.

[76] B. M. Mohan and S. K. Kar, "State estimation using shifted Legendre polynomials," *Third Int. Conf. on Industrial and Information Systems (ICIIS)*, Kharagpur, INDIA, pp: 1-6, December 8-10, 2008.

[77] B. M. Mohan and S. K. Kar, "Block-pulse functions approach to analysis of linear optimal control systems incorporating observers," *Third Int. Conf. on Industrial and Information*

Systems (ICIIS), Kharagpur, INDIA, pp: 1-4, December 8-10, 2008.

[78] B. M. Mohan and S. K. Kar, "State estimation using Block-Pulse Functions," *IEEE INDICON*, Kanpur, INDIA, pp: 280-285, December 11-13, 2008.

[79] B. M. Mohan and S. K. Kar, "Shifted Legendre polynomials approach to analysis of linear optimal control systems incorporating observers," *IEEE INDICON*, Kanpur, INDIA, pp: 383-387, December 11-13, 2008.

[80] S. K. Kar, "Orthogonal functions approach to optimal control of linear time-invariant systems described by integro-differential equations," *KLEKTRIKA*, vol. 11, no 1, pp: 15-18, 2009.

[81] F. Khellat, "Optimal control of linear time-delayed systems by linear Legendre multiwavelets," *J. Optimization Theory and Application*, vol. 143, no. 1, pp: 107-121, 2009.

[82] R. Ebrahimi, M. A. Vali, M. Samavat and A. A. Gharavisi, "A computational method for solving optimal control of singular systems using the Legendre wavelets," *ICGST-ACSE Journal*, vol. 9, no. 2, pp: 1-6, 2009.

[83] M. Razzaghi, "Optimization of time delay systems by hybrid functions," *Optimization and Engineering*, vol. 10, no. 3, pp: 363-376, 2009.

[84] X. T. Wang, "A numerical approach of optimal control for generalized delay systems by general Legendre wavelets," *Int. J. Computer Mathematics*, vol. 86, no. 4, pp: 743-752, 2009.

[85] B. M. Mohan and S. K. Kar, "Optimal control of singular systems via block-pulse functions," *Int. Conf. on Recent Advances in Mathematical Sciences and Applications*, Gayatri Vidya Parishad, Visakhapatnam, INDIA, December 19-22, 2009.

[86] B. M. Mohan and S. K. Kar, "Estimation and optimal control with reduced order observers via orthogonal functions," *Int. J. Signal, System, Control and Engineering Application*, vol. 3, no. 1, pp: 1-6, 2010.

[87] B. M. Mohan and S. K. Kar, "Optimal control of multi-delay systems via orthogonal functions," *Int. J. Advanced Research in Engineering and Technology*, vol. 1, no. 1, pp: 1-24, 2010.

[88] H. R. Marzban and M. Razzaghi, "Rationalized Haar approach for nonlinear constrained optimal control problems," *Applied Mathematical Modelling*, vol. 34, no. 1, pp: 174-183, 2010.

[89] B. M. Mohan and S. K. Kar, "State estimation using orthogonal functions," *Int. J. Mathematics and Engineering with Computers*, vol. 1, no. 1-2, pp: 1-20, 2010.

[90] B. M. Mohan and S. K. Kar, "Optimal control of multi-delay systems via block-pulse functions," *Fifth Int. Conf. on Industrial and Information Systems (ICIIS)*, Mangalore, INDIA, pp: 614-619, July 29-August 1, 2010.

[91] B. M. Mohan and S. K. Kar, "Orthogonal functions approach to optimal control of delay systems with reverse time terms," *J. The Franklin Institute*, vol. 347, no. 9, pp: 1723-1739, 2010.

[92] B. M. Mohan and S. K. Kar, "Optimal control of singular systems via orthogonal functions," *Int. J. Control, Automation and Systems*, vol. 9, no. 1, pp: 145-152, 2011.

[93] H. Ma, T. Qin and W. Zhang, "An efficient Chebyshev algorithm for the solution of optimal control problems," *IEEE Trans. Automatic Control*, vol. 56, no. 3, pp: 675-680, 2011.

[94] B. M. Mohan and S. K. Kar, "Optimal control of multi-delay systems via shifted Legendre polynomials," *Int. Conf. on Energy, Automation and Signals (ICEAS)*, Bhubaneswar, INDIA, December 28-30, 2011.

[95] B. M. Mohan and S. K. Kar, "Optimal control of nonlinear systems via orthogonal functions," *Int. Conf. on Energy, Automation and Signals (ICEAS)*, Bhubaneswar, INDIA, December 28-30, 2011.

Index